Discrete dynamical systems and chaos

Pitman Monographs and
Surveys in Pure and Applied Mathematics 62

Discrete dynamical systems and chaos

Mario Martelli

California State University

Longman
Scientific &
Technical

Copublished in the United States with
John Wiley & Sons, Inc., New York

Longman Scientific & Technical
Longman Group UK Limited
Longman House, Burnt Mill, Harlow
Essex CM20 2JE, England
and Associated companies throughout the world.

Copublished in the United States with
John Wiley & Sons Inc., 605 Third Avenue, New York, NY 10158

First published 1992

AMS Subject Classification: 34, 39, 58F

ISSN 0269-3666

ISBN 0 582 09637 5

British Library Cataloguing in Publication Data

A catalogue record for this book is
available from the British Library

Library of Congress Cataloging-in-Publication Data

Discrete dynamical systems and chaos.
p. cm.
1. Differentiable dynamical systems. 2. Chaotic behavior in systems.
QA614.8.D57 1992
515'.352--dc20 92-30845
CIP

Printed and bound in Great Britain at the Bath Press, Avon

To my wife and my children

Contents

PREFACE

The purpose of this book is to bring some of the fundamental ideas on Discrete Dynamical Systems and Chaos at the level of those undergraduates, usually in their junior year, who have completed the standard Calculus sequence, which includes the study of the basic facts about Functions of Several Variables and elements of Linear Algebra. I feel that this is the best moment for them to be exposed, during their college training, to the new ideas and developments that Dynamical Systems and Chaos have generated during the last thirty years. As R. Devaney writes in his book An Introduction to Chaotic Dynamical Systems "*the field of dynamical systems and especially the study of chaotic systems has been hailed as one of the important breakthroughs in science in this century.*"

The students' degree of sophistication permits the presentation of a broad range of topics and a fairly deep analysis of some non trivial and historically interesting models.

The book is divided into seven chapters, and its content can be comfortably covered during a one-semester course, particularly if the teacher is satisfied with providing detailed proofs of only the fundamental results.

The first chapter presents definitions and general ideas about discrete dynamical systems, together with some examples of significant interest.

In section 1 we introduce standard definitions and terminology to be used throughout the book. In section 2 we give examples of systems, such as the system proposed by E.N. Lorenz to model atmospheric changes and the system proposed by J.J. Hopfield to model neural networks, which are later (Chapter 7) studied using the theory developed in the course .

The second chapter contains an extensive analysis of one-dimensional dynamical systems depending on one parameter. In the first section we introduce the cobweb method and the idea of conjugacy. In the second section we study the stability and instability of stationary states and periodic orbits. Section 3 presents an important theorem of A.N. Sarkovskii and a sufficient condition for all orbits of a dynamical system in **R** to be convergent to a stationary state. The last section is reserved to bifurcation points and bifurcation diagrams for one-parameter families of maps.

The purpose of this chapter is to give the students something to work on at the outset. Its only prerequisites are a few fundamental results from Calculus of One Variable.

The third chapter contains an overview of the topics on Linear Algebra and Calculus of Several Variables which are needed in the sequel. Many of the definitions and results have probably been encountered more than once. Therefore, details are kept to a minimum, and proofs are usually omitted.

In the first section one finds the standard definition of $\mathbf{R}^q$ and of its structure. Section 2 is devoted to matrices, with particular attention to eigenvalues and spectral radius. In the third section we present matrices as operators, we introduce the operator norm of a matrix, its relation to the spectral radius, the eigenvectors and the real canonical form of a matrix. Section 4 is a brief summary on differentiability and first order approximation.

In the fourth chapter we analyze discrete systems governed by the action of a matrix. Our study is based on three fundamental tools: the spectrum, the real canonical form of a matrix, and a fundamental property of the spectral radius.

The first section explores the relation between linear and affine systems using the idea of conjugacy. In section 2 we study the case of matrices having spectral radius smaller than 1, or such that the inverse has spectral radius smaller than 1. In section 3 we investigate the saddle case, namely the action of matrices with some eigenvalues in modulus smaller than one and the others in modulus larger than one. Finally, in section 4 we analyze the case when at least one of the eigenvalues has modulus one.

With chapter five we enter into the more challenging part of the book. In the first section we analyze systems with a unique equilibrium point and such that every orbit converges to it. Three types of maps are studied : contractive, gradient and triangular. Section 2 is devoted to local asymptotic stability of equilibrium points and periodic orbits. The third section deals with repellers and saddles, with a brief excursion on stable and unstable manifolds. In the fourth section we introduce two families of maps (dissipative and quasi-bounded) for which the existence of a bounded and invariant region can be easily established. We show here that the map proposed by Lorenz is of the first type, and the one used by Hopfield is of the second type. Section 5 presents three fundamental results on bifurcation.

Chapter six is devoted to chaotic behavior. The first section opens the chapter with the definition of the core of a dynamical system and of its attractors. In section 2 we present a definition of chaotic dynamical systems based on the presence of dense orbits and on the instability of all orbits. Sensitivity with respect to initial conditions and other alternative definitions of chaos are also presented. In section 3 we analyze the attractors of a chaotic system from the point of view of their dimension. Two types of dimension are discussed, the so-called capacity which is strictly related to the Hausdorff dimension and the correlation dimension of Grassberger-Procaccia. In the last section Lyapunov exponents are discussed together with their relation to stability and sensitivity with respect to initial conditions.

In the last chapter we present an extensive, although not complete, analysis of the models introduced in the first chapter, namely a blood cell population model, the predator-prey model for competition between two species, the model proposed by Lorenz as an approximation to the dynamics of atmospheric changes, and the Hopfield model of a neural network.

At the end of each chapter the reader finds some problems numbered as the corresponding sections of the chapter. Some of the problems can be solved without any help from a computing device, the answers to others are simpler to find with the aid of a symbolic manipulator like Mathematica, and a few cannot be done without a computer, or a programmable calculator.

An extensive analysis of a dynamical system is rarely possible without the significant help that can be provided by a computer, particularly when combined with at least a working knowledge of a powerful symbolic manipulator. It is hard, and maybe even impossible, to get hands-on experience without using the enormous computing capability of a machine. Students are urged to learn how to use at least one of the many available programs. Some are strictly designed for the study of dynamical systems. Others address a much broader range of topics, and have quite a few features that can be exploited for a successful numerical study of discrete dynamical systems. Some suggestions on programs and books are listed before the index.

Many scientists have written extensively about the importance of the topics presented in this book. It is hard to add something new to the things they have said so beautifully and appropriately. In particular, I believe that the number of mathematicians who feel that these topics should be made accessible to our undergraduates largely exceeds the number of those who are still reluctant to "follow the trend." I belong to the first group, and this book is my attempt to provide one more tool for reaching the goal. Many friends have helped me in different ways. I am thankful to all of them. In particular I would like to mention Alfonso Albano, Stavros Busenberg, Courtney Coleman, Massimo Furi, and William Gearhart. When I was uncertain about the presentation of certain topics, or the most appropriate examples for clarifying definitions and theorems, they provided invaluable help. I am indebted to my students, especially W. Buckley and D. Fieri, and to my daughter Monica for finding numerous errors and misprints. I am thankful to my Department, and particularly to the chairman Jim Friel, for encouraging me in this enterprise by making a course on Discrete Dynamical Systems and Chaos mandatory for all math majors at CSUF.

Last, but not least, I would like to express my appreciation to the publisher for making the results of my effort available to the community of scientists and teachers. I sincerely hope they will find this book useful.

Mario Martelli

Claremont, 1992

The book has been written using Microsoft Word 4.0 on a Macintosh.
The graphs have been produced using Mathematica.
Some programs have been contributed by two students : Eric Ramberg and Ravi Montenegro.

CHAPTER 1

DISCRETE DYNAMICAL SYSTEMS

SUMMARY

The purpose of this chapter is to provide a general description of discrete dynamical systems and their fundamental features (section 1), and to present some interesting examples of systems derived from the current research literature (section 2).

Section 1. Generalities on Dynamical Systems

1. Dynamical processes

The fundamental component of any mathematical model of a dynamical process, be it chemical, biological, economic or otherwise, is a set of equations or equalities relating the variables and parameters which describe the state of the process. In the selection of the variables the researcher always strikes a compromise between the complexity of the system and an objective representation of those features which are perceived as necessary to its useful analysis. Parameters are usually included with the purpose of representing control mechanisms of the process and to curve fit the model to experimental data.

As a rule of thumb, one can say that there is an optimal range for the level of complexity of any model of a dynamical process. Below this range, the validity of the model is strongly biased toward the theoretical side, and above this range the model loses its synthetic advantage.

The existence of this optimal range does not usually help in finding it. The two extrema of the spectrum are in general well delimited, but large areas of uncertainty remain. Different models of the same process may fall into this area and have different merits. It may be very difficult, or even impossible, to design a "universal" model of the process, namely a model in which all advantages and key features are incorporated.

From the point of view of the time variable, which is always included, either explicitly or implicitly, in any model of a dynamical process, we can distinguish between

- *short term or long term behavior;*
- *continuous or discrete models;*
- *presence or absence of delayed effects.*

In this book we are mainly interested in the long term behavior, and we study discrete models.

The choice of the time unit is not a trivial issue, but will not be addressed here.

We assume time intervals to be equally spaced and we denote them by 0, 1, 2, ... The state variables are named $x_1, x_2, ..., x_q$, and the parameters $a_1, a_2, ..., a_m$. The state of the system at time $n+1$ is assumed to be determined by the state at time n according to a vector equation of the form

$$(1.1.1) \qquad \mathbf{x}_{n+1} = F(\mathbf{x}_n, \mathbf{a})$$

where

- $\mathbf{x}_n = (x_1, x_2, ..., x_q)_n$ is the state of the q variables of the system at time n ;
- $\mathbf{a} = (a_1, a_2, ..., a_m)$ is the vector whose components are the m parameters of the system;
- $F(\mathbf{x}, \mathbf{a}) = (F_1(\mathbf{x}, \mathbf{a}), F_2(\mathbf{x}, \mathbf{a}), ..., F_q(\mathbf{x}, \mathbf{a}))$ are the q functions governing the process.

Example 1.1.1 Assume that the value x_{n+1} of x at time $n+1$, depends on the value x_n at time n, according to the relation

$$(1.1.2) \qquad x_{n+1} = ax_n(1 - x_n), \qquad x \geq 0,$$

where $a \in [0, 4]$ is a control parameter. Then $q = 1, m = 1, F(x, a) = ax(1 - x)$. The dynamical system (1.1.2) is usually called *logistic*.

Example 1.1.2 Assume that in a two species population of prey and predators the size of the two species at time $n+1$ depends on their size at time n according to the relations

$$(1.1.3) \qquad \begin{cases} x_{n+1} = ax_n - bx_ny_n \\ y_{n+1} = cy_n + dx_ny_n \end{cases}$$

where a, b, c, d are non-negative numbers, $a > 1$, $c < 1$, and x, y denote the prey and predators population respectively. Then $\mathbf{x} = (x, y) \in \mathbf{R}^2$, $\mathbf{a} = (a, b, c, d) \in \mathbf{R}^4$.
The function

$$F : \mathbf{R}^2 \to \mathbf{R}^2 \quad \text{is} \quad F(\mathbf{x}, \mathbf{a}) = F((x, y), (a, b, c, d)) = (F_1(\mathbf{x}, \mathbf{a}), F_2(\mathbf{x}, \mathbf{a})),$$

where $F_1(\mathbf{x}, \mathbf{a}) = ax - bxy$ and $F_2(\mathbf{x}, \mathbf{a}) = cy + dxy$.

The presence of delayed effects on a discrete dynamical system can be eliminated by embedding it into a higher-dimensional process. For example, the discrete system

$$(1.1.4) \qquad \mathbf{x}_{n+1} = F(\mathbf{x}_n, \mathbf{x}_{n-1}, \mathbf{a})$$

where a delay of one-time unit directly affects the state at time n+1, can be rewritten in the following form:

$$(1.1.5) \qquad \begin{cases} \mathbf{x}_{n+1} = F(\mathbf{x}_n, \mathbf{y}_n, \mathbf{a}) \\ \mathbf{y}_{n+1} = \mathbf{x}_n . \end{cases}$$

Equivalently,

$$(1.1.6) \qquad \mathbf{z}_{n+1} = G(\mathbf{z}_n, \mathbf{a})$$

where $\mathbf{z}_n = (\mathbf{x}_n, \mathbf{y}_n)$, $G(\mathbf{x}_n, \mathbf{y}_n) = (F(\mathbf{x}_n, \mathbf{y}_n, \mathbf{a}), \mathbf{x}_n)$.

Example 1.1.3 Let $x_{n+1} = ax_n(1 - x_{n-1})$. This one-dimensional dynamical system, containing a delay of a one-time unit, can be embedded into a two-dimensional system with no delay by setting $y_{n+1} = x_n$. We obtain the system

$$\begin{cases} x_{n+1} = ax_n(1 - y_n) \\ y_{n+1} = x_n . \end{cases}$$

Example 1.1.4 Let $x_{n+1} = a\, x_n(1- x_{n-2})$. The dynamical system contains a delay of two-time units. We can embed it into a three-dimensional system with no delay. Set $z_{n+1} = y_n$ and $y_{n+1} = x_n$. We obtain

$$\begin{cases} x_{n+1} = ax_n(1 - z_n) \\ y_{n+1} = x_n \\ z_{n+1} = y_n . \end{cases}$$

In this book we always assume that $\mathbf{x}_{n+1}$ depends directly only on $\mathbf{x}_n$. Systems with delay, i.e. when $\mathbf{x}_{n+1}$ depends directly on one or more states of the form $\mathbf{x}_{n-k}$, $k \geq 1$, are considered embedded into higher-dimensional systems with no delay.

Given an initial state $\mathbf{x}_0$, the evolution of the system starting from $\mathbf{x}_0$ is determined by the sequence of states

$$(1.1.7) \qquad \mathbf{x}_0,\ \mathbf{x}_1 = F(\mathbf{x}_0, \mathbf{a}),\ \mathbf{x}_2 = F(\mathbf{x}_1, \mathbf{a}) = F^2(\mathbf{x}_0, \mathbf{a}), \ldots\ .$$

The sequence is denoted by $O(\mathbf{x}_0, \mathbf{a})$ and it is called the **orbit** or **trajectory** of the system starting from $\mathbf{x}_0$ and with parameter $\mathbf{a}$. The control parameters are normally held constant, but this limitation may need to be lifted when real processes are studied. In this case the state variables and the control parameters must be clearly identified, and how and when the control parameters change must be specified.

In this book the analysis of dynamical systems will be focused on two different, but strictly related issues:

1) behavior of $F(., \mathbf{a})$ for $\mathbf{a}$ fixed;
2) changes in this behavior as $\mathbf{a}$ is changed.

1) A dynamical system is given when the value of the vector parameter $\mathbf{a}$ is selected. Therefore, when we talk about a discrete dynamical system we should consider $F(\mathbf{x}, \mathbf{a})$ as a function of the q state variables $(x_1, x_2, \ldots, x_q) = \mathbf{x}$ and we should view the m parameters $(a_1, a_2, \ldots, a_m) = \mathbf{a}$ as constants. When these constant values are clear from the context, the map $F(\mathbf{x}, \mathbf{a})$ may be indicated simply by $F(\mathbf{x})$. At least for illustration purposes we sometimes present dynamical systems with only state variables.

The main object of our analysis in this stage of the study is the behavior of the orbits of $F(., \mathbf{a})$ for different choices of the initial condition $\mathbf{x}_0$.
The orbit's notation $O(\mathbf{x}_0, \mathbf{a})$ is frequently changed into $O(\mathbf{x}_0)$.

2) Our second goal is to study the changes in the characteristics and behavior of $F(., \mathbf{a})$ as the parameter $\mathbf{a}$ is changed. In the case when $\mathbf{a}$ is a scalar (denoted by a) the set of functions $\{F(., a) : a \in \mathbf{R}\}$ is called <u>one-parameter</u> family of maps. Our study is focused on the properties of such families.

Example 1.1.5 Let $F(x, a, b) = ax(b - x)$.
For a, b fixed the map F defines a discrete dynamical system in $\mathbf{R}$. Given $x_0 \in \mathbf{R}$ we have

$$x_1 = ax_0(b - x_0),\ \ x_2 = ax_1(b - x_1),\ \ldots\ .$$

The parameters a, b are considered constants even if no specific numerical value has been selected. Now, suppose $b = 1$. We obtain

$$F(x, a, 1) = ax(1 - x) = F(x, a).$$

The set of functions $\{F(., a) : a \in \mathbf{R}\}$ is a one-parameter family of maps. To illustrate the type of questions we will investigate later, notice that $F(0.5, 0) = 0$, $F(0.5, 2) = 0.5$. The two orbits are different.

$$O(0.5, 0) \text{ is } x_0 = 0.5, x_1 = x_2 = ... = 0,$$

while

$$O(0.5, 2) \text{ is } x_0 = x_1 = x_2 = ... = 0.5 .$$

2. Stationary states

A point $\mathbf{x}_s$ is called an **equilibrium point**, or a **fixed point**, or a **stationary state** of the system with parameter $\mathbf{a}$ if

$$\mathbf{x}_s = F(\mathbf{x}_s, \mathbf{a}). \tag{1.1.8}$$

Example 1.1.6 A stationary state of the system $x_{n+1} = ax_n(1 - x_n)$ must satisfy the equation $x = ax(1 - x)$. We see that $x_s = 0$ is a fixed point regardless of the value of a. Assuming $x \neq 0$, we obtain

$$1 = a(1 - x) \qquad \text{or} \qquad ax = a - 1. \tag{1.1.9}$$

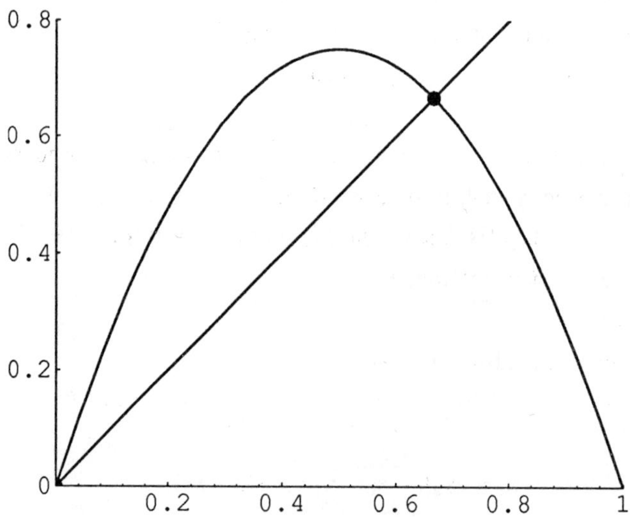

Fig. 1.1.1 The equilibrium points of the dynamical system governed by the function $F(x) = 3x(1 - x)$ are the intersections of the graph of F with the line $y = x$.

Since x must be positive (see Example 1.1.1), a second stationary state $x_s = (a-1)/a$ appears when $a > 1$.

Example 1.1.7 The equilibrium points of the predator-prey system of Example 1.1.2 are the solutions of the equations

$$\begin{cases} x = ax - bxy \\ y = cy + dxy . \end{cases}$$

The point $(0, 0)$ is a solution regardless of the values of a, b, c, d. From the inequalities $0 < c < 1$ and $a > 1$, we obtain that if $x = 0$ then $y = 0$ and vice versa. Thus, the search of additional stationary states can assume $xy > 0$. We obtain the system

$$\text{(1.1.10)} \qquad \begin{cases} 1 = a - by \\ 1 = c + dx \end{cases}$$

which gives the equilibrium point $\mathbf{x}_s = (\frac{1-c}{d}, \frac{a-1}{b})$.

Notice that the orbit starting from a stationary state $\mathbf{x}_s$ will not move away from it. This situation is possible in theory, but not in practice, since every process is subject to small fluctuations which are normally not accounted for in a model. These fluctuations will drive the systems away from the equilibrium point. This observation, together with other considerations which will be discussed later, motivates the introduction of the idea of **stability** and of its analysis. Some fundamental definitions are presented in this chapter. Other concepts related to stability will be discussed in later chapters.

3. Stability of stationary states

An equilibrium point $\mathbf{x}_s$ is stable if all future states of every orbit starting at a point $\mathbf{x}_0$ close to $\mathbf{x}_s$ remain close to $\mathbf{x}_s$. More precisely, stability can be defined as follows.

Definition of stable equilibrium

An equilibrium point $\mathbf{x}_s$ is said to be **stable** if for every $r > 0$ there exists a positive number d such that

$$\text{(1.1.11)} \qquad \|\mathbf{x}_0 - \mathbf{x}_s\| \leq d \text{ implies } \|\mathbf{x}_n - \mathbf{x}_s\| \leq r \text{ for all } n \geq 1.$$

The symbol $\|\mathbf{x}\|$ denotes the so-called "Euclidean norm" of $\mathbf{x}$, which is defined as

(1.1.12) $\|\mathbf{x}\| = \|(x_1, ..., x_q)\| = \sqrt{x_1^2 + ... + x_q^2}$.

Different norms can be used in (1.1.11), like the "taxi-cab norm"

(1.1.13) $\|\mathbf{x}\|_{TC} = |x_1| + ... + |x_q|$.

The taxi-cab and the Euclidean norm are related by the inequalities

(1.1.14) $q^{-.5}\|\mathbf{x}\|_{TC} \le \|\mathbf{x}\| \le \|\mathbf{x}\|_{TC}$.

In general, given any norm $\|.\|_o$ in $\mathbf{R}^q$ (see Chapter 3) we can find two positive constants $m < M$, such that

(1.1.15) $m\,\|\mathbf{x}\|_o \le \|\mathbf{x}\| \le M\,\|\mathbf{x}\|_o$.

Therefore, the definition of stability is independent of the norm. The same is true for all definitions and results contained in this book. They are stated, with few exceptions, in terms of the Euclidean norm, but they are valid, with suitable minor adjustments, in any other norm.

Example 1.1.8 Let $F(x) = 1 - x$. The point $x_s = 0.5$ is the only fixed point of F. For every other initial state x_0 we have $F(x_0) = 1 - x_0$, $F(F(x_0)) = F(1 - x_0) = x_0$. Thus, the orbit $O(x_0)$ does not converge to 0.5, but it remains always at the same distance from it. The equilibrium point $x_s = 0.5$ is stable with $d = r$.

Example 1.1.9 Let $F(\mathbf{x}) = F(x, y) = (x\text{Cos}\,\theta + y\text{Sin}\,\theta, - x\text{Sin}\,\theta + y\text{Cos}\,\theta)$, $\theta \in [0, 2\pi)$. Notice that $\mathbf{0} = (0, 0)$ is an equilibrium point of F. Using the Euclidean norm we have

$$\|F(\mathbf{x})\|^2 = (x\text{Cos}\,\theta + y\text{Sin}\,\theta)^2 + (- x\text{Sin}\,\theta + y\text{Cos}\,\theta)^2 = x^2 + y^2 = \|\mathbf{x}\|^2 .$$

All states $\mathbf{x}_n$, $n = 0, 1, ...$ of an orbit $O(\mathbf{x}_0)$ are at the same distance from $\mathbf{0}$, $\|\mathbf{x}_n\| = \|\mathbf{x}_0\|$. The origin is stable with $d = r$.
Using a different norm we may have to change the choice of d, but $\mathbf{0}$ will remain a stable equilibrium.

In some cases all orbits starting from points close to $\mathbf{x}_s$ not only remain close to $\mathbf{x}_s$, but they converge to it. This situation will be analyzed in later chapters. In the meantime we illustrate it with an example.

Example 1.1.10 Let $F(x, a) = ax(1 - x)$, $x \in [0, 1]$. Then $F(0) = F(1) = 0$. We want to show that every orbit will converge to 0 provided that $a \in [0, 1)$.
First observe that $0 \leq 1 - x \leq 1$. This implies

$$0 \leq x_1 = ax_0(1 - x_0) \leq ax_0, \quad 0 \leq x_2 = ax_1(1 - x_1) \leq ax_1 \leq a^2x_0.$$

In general

$$0 \leq x_n = ax_{n-1}(1 - x_{n-1}) \leq ax_{n-1} \leq \ldots \leq a^n x_0.$$

Since $a \in [0, 1)$ we obtain $a^n \to 0$ as $n \to \infty$.

Definition of unstable equilibrium

A stationary state $\mathbf{x}_s$ which is not **stable** is said to be **unstable**. In other words the instability of $\mathbf{x}_s$ is equivalent to the existence of a positive number r, such that for every $d > 0$ there exists a point $\mathbf{x}_0$, which is not further than d from $\mathbf{x}_s$, $\|\mathbf{x}_0 - \mathbf{x}_s\| \leq d$, but whose trajectory goes further than r from $\mathbf{x}_s$, i.e. $\|\mathbf{x}_n - \mathbf{x}_s\| > r$, for some positive integer n.

Example 1.1.11 Let $F : \mathbf{R} \to \mathbf{R}$ be defined by

$$F(x) = \begin{cases} 0.5x & \text{for } x \leq 0 \\ 2x & \text{for } x > 0 . \end{cases}$$

Every orbit starting at a point to the left of the unique stationary state $x_s = 0$ will converge to 0, while every orbit starting at a point to the right of 0 will go to infinity. Thus, 0 is an unstable stationary state. We could choose $r = 1$ in the definition of instability. Then, for every $d > 0$ we can select any point $x_0 \in (0, d)$. Since the orbit $O(x_0)$ goes to infinity there exists n such that $x_n > 1$.

Notice that the instability of a stationary state $\mathbf{x}_s$ does not require all orbits starting from nearby points to eventually move away from $\mathbf{x}_s$, but only that we can find points with this property as close as we like to $\mathbf{x}_s$. However, there are stationary states having the property that every orbit starting from points close to them will move away. This situation will be discussed in later chapters. For the moment we illustrate it with an example.

Example 1.1.12 Let $F(x) = x + x^3$. The only fixed point of F is $x_s = 0$. Moreover, for every $x_0 > 0$, we have $x_0 < x_1 < ... < x_{n-1} < x_n ...$ and the sequence of iterates goes to $+\infty$. Similarly, for every $x_0 < 0$, we have $x_0 > x_1 > ... > x_{n-1} > x_n ...$ and the sequence of iterates goes to $-\infty$.

4. Periodic orbits

An orbit $O(\mathbf{x}_0, \mathbf{a})$ is said to be periodic of period p if

(1.1.16) $$\mathbf{x}_p = \mathbf{x}_0 .$$

The smallest integer p such that (1.1.16) holds is called the **prime period** (or simply the **period**) of the orbit, which is denoted by $O(\mathbf{x}_0, \mathbf{a}, p)$. The choice of the attribute "prime" is motivated by the fact that the orbit cannot be decomposed into smaller loops. Sometimes we say that $\mathbf{x}_0$ is a periodic point of period p. A stationary state can be regarded as a periodic point of period 1.

Notice that every point $\mathbf{x}_i$, $i = 0, 1, ..., p-1$, of the orbit $O(\mathbf{x}_0, \mathbf{a}, p)$ is also periodic of the same period p. Thus $O(\mathbf{x}_0, \mathbf{a}, p)$ contains exactly p distinct periodic points of period p.

Example 1.1.13 Consider the dynamical system

$$F(x, a) = ax(1 - x) , a \in [0, 4].$$

A point x is periodic of prime period 2 if

$$F(x, a) \neq x \text{ and } F(F(x, a), a) = F^2(x, a) = x.$$

Thus the periodic points of prime period 2 satisfy the equality

$$a(ax(1 - x))(1 - ax(1- x)) = x.$$

We must have $x \neq 0$ since 0 is a fixed point. Therefore

$$a^2(1 - x)(1 - ax(1 - x)) = 1, \quad \text{or} \quad a^2(1 - x) - a^3x(1 - x)^2 = 1.$$

Moreover $x \neq \frac{a-1}{a}$ since $\frac{a-1}{a}$ is a stationary state. Hence the periodic points of prime period 2 are the real solutions of the quadratic equation

$$a^2x^2 - (a^2 + a)x + a + 1 = 0.$$

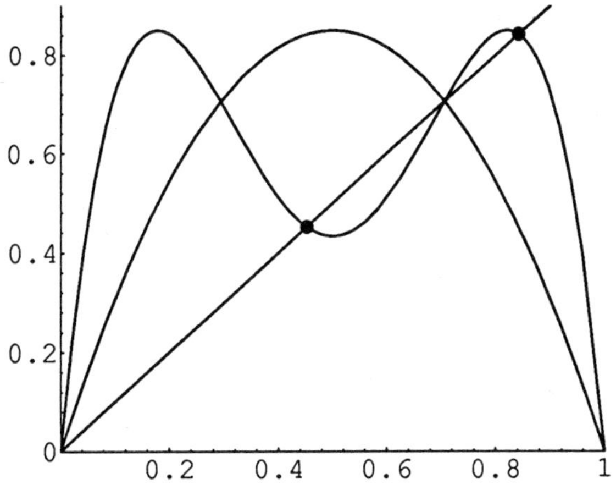

Fig. 1.1.2 The periodic points of prime period 2 of $F(x) = 3.4x(1 - x)$ are the intersections of the graph of $F^2(x) = F(F(x))$ with the line $y = x$, at those points where $F(x) \neq x$.

No periodic orbit of period 2 is possible if $a < 3$ since the discriminant of the quadratic equation is negative. At $a = 3$ the equation has the only solution $\frac{2}{3}$ which is a stationary state of the system. For $a > 3$ there are *exactly two* periodic points of period 2, namely

$$x_1 = \frac{a + 1 + \sqrt{a^2 - 2a - 3}}{2a}, \quad x_2 = \frac{a + 1 - \sqrt{a^2 - 2a - 3}}{2a}.$$

Hence, they belong to the same orbit, i.e. $F(x_1, a) = x_2$, $F(x_2, a) = x_1$.

Every point $\mathbf{x}_i$ of the periodic orbit $O(\mathbf{x}_0, \mathbf{a}, p)$ is a stationary state of the map $F^p(\mathbf{x}, \mathbf{a})$ which is obtained as composition of the map $F(\mathbf{x}, \mathbf{a})$ with itself p-times,

$$(1.1.17) \qquad \mathbf{x}_i = F^p(\mathbf{x}_i, \mathbf{a}), \qquad i = 0, 1, 2, ..., p-1.$$

Example 1.1.14 From Example 1.1.13 we know that $F(x_1, a) = x_2$ and $F(x_2, a) = x_1$. Hence $F^2(x_1, a) = F(F(x_1, a), a) = x_1$ and $F^2(x_2, a) = F(F(x_2, a), a) = x_2$. For example, for $a = 4$ we obtain

$$x_1 = \frac{5 + \sqrt{5}}{8} \text{ and } x_2 = \frac{5 - \sqrt{5}}{8}.$$

We can easily verify that

$$F^2(\frac{5 + \sqrt{5}}{8}, 4) = F(\frac{5 - \sqrt{5}}{8}, 4) = \frac{5 + \sqrt{5}}{8},$$

$$F^2(\frac{5-\sqrt{5}}{8}, 4) = F(\frac{5+\sqrt{5}}{8}, 4) = \frac{5-\sqrt{5}}{8}.$$

5. Stability and instability of periodic orbits

Recall that each state $\mathbf{x}_i$ of the periodic orbit of prime period p,

$$O(\mathbf{x}_0, p) = \{ \mathbf{x}_0, \mathbf{x}_1, ..., \mathbf{x}_{p-1}, \mathbf{x}_p = \mathbf{x}_0, \mathbf{x}_1, ...\},$$

is an equilibrium point of the map F^p, i.e. $F^p(\mathbf{x}_i, \mathbf{a}) = \mathbf{x}_i$, i = 0, 1, ..., p-1. We say that the orbit $O(\mathbf{x}_0, \mathbf{a}, p)$ is **stable (unstable)** if each point $\mathbf{x}_i$ is a stable (unstable) stationary state of the dynamical system governed by F^p.

Example 1.1.15 Let F(x) = 1- x. Every point $x \neq 0.5$ is periodic of prime period 2. For example F(0) = 1 and F(1) = 0. The periodic orbit {0, 1} is stable. In fact, F^2 is the identity map for which each point is a stationary stable state.

Example 1.1.16 Let $F(\mathbf{x}) = F(x, y) = (-y, x)$. Every point $\mathbf{x} \neq \mathbf{0}$ is periodic of period 4. For example from $\mathbf{x}_0 = (1, 0)$ we obtain

$$\mathbf{x}_1 = (0, 1),\ \mathbf{x}_2 = (-1, 0),\ \mathbf{x}_3 = (0, -1),\ \mathbf{x}_4 = \mathbf{x}_0, \dots .$$

The orbit is stable, since F^4 is the identity map for which each point is a stationary stable state.

The definition of stability for stationary states and periodic orbits can be extended to any orbit $O(\mathbf{x}_0)$ by requiring that for every $r > 0$ there exists $d > 0$ such that $\|\mathbf{x}_0 - \mathbf{y}_0\| \leq d$ implies $\|\mathbf{x}_n - \mathbf{y}_n\| \leq r$ for every $n \geq 1$. This extension and its implications will be discussed in Chapter 6. In the meantime we present an example of a stable orbit which is not periodic.

Example 1.1.17 Let $F(\mathbf{x}) = F(x, y) = (x\text{Cos}1 - y\text{Sin }1, x\text{Sin }1 + y\text{Cos }1)$. The dynamical system generated by F in $\mathbf{R}^2$ is a rotation of an angle $\theta = 1$ radian. Consider the initial condition $\mathbf{x}_0 = (1, 0)$. We want to show that the orbit O((1, 0)) is not periodic. Assume on the contrary that the orbit is periodic of prime period p. Since

$$F^n(x, y) = (x\text{Cos } n - y\text{Sin } n, x\text{Sin } n + y\text{Cos } n),$$

we obtain

$$F^p(1, 0) = (\text{Cos}\, p, \text{Sin}\, p) = (1, 0),$$

which implies $p = 2n\pi$. This equality is impossible since $\frac{p}{2n}$ is rational and π is not. The orbit $O(\mathbf{x}_0)$ is stable since

$$\|\mathbf{x}_n - \mathbf{y}_n\| = \|F^n(\mathbf{x}_0) - F^n(\mathbf{y}_0)\| = \|\mathbf{x}_0 - \mathbf{y}_0\|,$$

an equality which can be easily established since F is a rotation (a rigid motion).

6. **Limit points, limit sets and aperiodic orbits**

The orbit $O(\mathbf{x}_0, \mathbf{a})$ of the dynamical system

(1.1.18) $\quad \mathbf{x}_{n+1} = F(\mathbf{x}_n, \mathbf{a})$

has infinitely many states, namely $\mathbf{x}_0, \mathbf{x}_1, \mathbf{x}_2, ..., \mathbf{x}_n, ...$. These states may not be all distinct. They are all equal when $\mathbf{x}_0 = \mathbf{x}_s$ is a stationary point; and there are only p distinct when the orbit $O(\mathbf{x}_0, \mathbf{a}, p)$ is periodic of prime period p.

Definition of limit point

A point $\mathbf{z}$ is said to be a **limit point** of $O(\mathbf{x}_0, \mathbf{a})$ if there exists a subsequence $\{\mathbf{x}_{n_k} : k = 0, 1, ...\}$ of $O(\mathbf{x}_0, \mathbf{a})$ which converges to $\mathbf{z}$, i.e. $\|\mathbf{x}_{n_k} - \mathbf{z}\| \to 0$ as $k \to \infty$. A stationary orbit has only one limit point: $\mathbf{x}_s$. A periodic orbit $O(\mathbf{x}_0, \mathbf{a}, p)$ has exactly p limit points: $\mathbf{x}_0, ..., \mathbf{x}_{p-1}$.

Example 1.1.18 In the system

$$x_{n+1} = x_n(1 - x_n)$$

the point 0 is the only limit point of the stationary orbit $O(0)$. It is also the only limit point of the orbit

$$x_0 = \frac{1}{2},\ x_1 = \frac{1}{4},\ x_2 = \frac{3}{16}, ...$$

Example 1.1.19 In the system $x_{n+1} = 1 - x_n$ the point $\frac{1}{4}$ is a limit point of the periodic orbit $x_0 = \frac{1}{4},\ x_1 = \frac{3}{4},\ x_2 = \frac{1}{4},\ ....$

Definition of limit set

The **limit set** of the orbit $O(\mathbf{x}_0, \mathbf{a})$ is the set of all limit points of the orbit.

We shall denote it by $L(\mathbf{x}_0, \mathbf{a})$, or simply $L(\mathbf{x}_0)$ whenever is convenient, and no confusion arises.

The limit set of an orbit $\{\mathbf{x}_0, \mathbf{x}_1, ...\}$ which converges to a stationary point $\mathbf{x}_s$, i.e. $\|\mathbf{x}_n - \mathbf{x}_s\| \to 0$ as $n \to \infty$, consists of the point $\mathbf{x}_s$. The limit set of an orbit which converges to a periodic orbit of period p consists of all points of the periodic orbit.

Some properties of the limit set of an orbit will be studied in Chapter 6. In the meantime we mention a fundamental equality between $L(\mathbf{x}_0)$ and its image under the function F, which generates the dynamical system :

$$(1.1.19) \qquad F(L(\mathbf{x}_0)) = L(\mathbf{x}_0)\,, \text{ with } F(L(\mathbf{x}_0)) = \{\, F(\mathbf{z}) \;: \mathbf{z} \in L(\mathbf{x}_0)\}.$$

Definition of asymptotically stationary or periodic orbit

An orbit $O(\mathbf{x}_0)$ is said to be **asymptotically stationary** if its limit set is a stationary state, and **asymptotically periodic** if its limit set is a periodic orbit. In the special case in which a state $\mathbf{x}_i$ of the orbit $O(\mathbf{x}_0)$ is a stationary state or a periodic point, we say that $O(\mathbf{x}_0)$ is respectively **eventually stationary** or **eventually periodic**.

Hence every eventually stationary or eventually periodic orbit is respectively asymptotically stationary or asymptotically periodic. The converse is not true. The orbit

$$x_0 = \frac{1}{2}\,,\ x_1 = \frac{1}{4}\,, x_2 = \frac{3}{16}\,, ...$$

of Example 1.1.18 is asymptotically stationary, but not eventually stationary.

Example 1.1.20 Let $B : [0, 1] \to [0, 1]$ be defined by $B(x) = 2x - [2x]$ where $[.]$ is the greatest integer function, namely the function which assigns to every $y \in \mathbf{R}$ the greatest integer less than or equal to y. Notice that 0 is a stationary state of B, and

$$B(\tfrac{3}{8}) = \frac{3}{4}\,,\ B(\frac{3}{4}) = \frac{1}{2}\,,\ B(\frac{1}{2}) = 0.$$

Hence the orbit

$$x_0 = 3/8\ ,\ x_1 = 3/4,\ x_2 = 1/2\ ,\ x_3 = x_4 = \dots = x_n = \dots = 0$$

is eventually stationary.

Similarly, the orbit

$$x_0 = \frac{1}{24},\ x_1 = \frac{1}{12},\ x_2 = \frac{1}{6},\ x_3 = \frac{1}{3},\ x_4 = \frac{2}{3},\ x_5 = \frac{1}{3},\ x_6 = \frac{2}{3}\ ,\ x_7 = \frac{1}{3},\ \text{etc.}$$

is eventually periodic.

<u>Definition of aperiodic orbit</u>

We say that the orbit $O(\mathbf{x}_0)$ is **aperiodic** if its limit set $L(\mathbf{x}_0)$ is not finite.

Example 1.1.21 Fig. 1.1.3a and Fig. 1.1.3b provide numerical evidence that the orbit of the dynamical system $K(x) = 4x(1 - x)$ starting from $x_0 = .3$ is aperiodic. Notice that $K(x)$ is the map $F(x, a) = ax(1 - x)$ for $a = 4$. To visualize the behavior of the orbit we plot (x_n, x_{n+1}) in Fig. 1.1.3a, and (x_n, x_n) in Fig. 1.1.3b.

It seems that the sequence of points $\{(x_n, x_{n+1}),\ n = 1, 2, \dots\}$ is able to "reconstruct" the entire arc of the parabola $4x(1 - x)$ in the interval $[0, 1]$ (Fig. 1.1.3a) and the orbit $O(.3)$ sweeps the entire interval $[0, 1]$ (Fig. 1.1.3b).

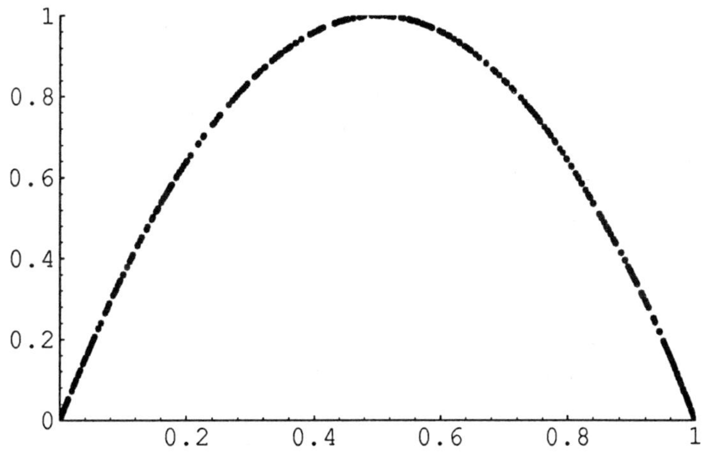

Fig. 1.1.3a The orbit of the system governed by $K(x) = 4x(1-x)$, starting at $x_0 = 0.3$ is plotted along the x-axis and the orbit starting from $x_1 = K(x_0)$ is plotted along the y-axis.
We obtain pairs of the form (x_n, x_{n+1}). It seems that the orbit is able to reconstruct the entire arc of the parabola.

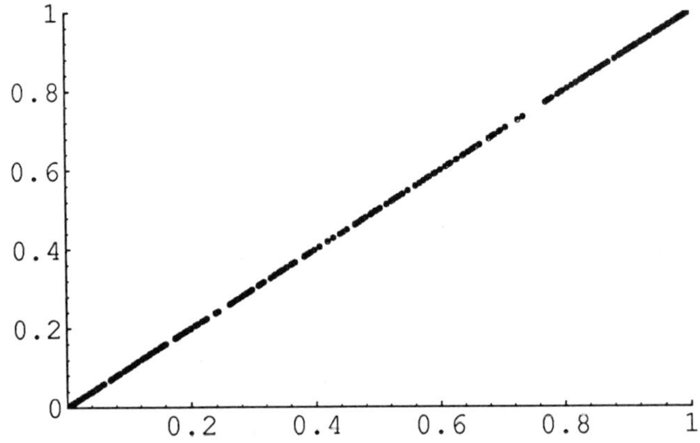

Fig. 1.1.3b The orbit of the system governed by $K(x) = 4x(1-x)$, starting at $x_0 = 0.3$ is plotted along the x-axis and along the y-axis, giving pairs of the form (x_n, x_n). The orbits "sweeps" the entire interval $[0, 1]$.

Example 1.1.22 Let $F(x, y) = (x\text{Cos}\,1 - y\text{Sin}\,1, x\text{Sin}\,1 + y\text{Cos}\,1)$ and $\mathbf{x}_0 = (1, 0)$. We have seen before that the orbit $O(\mathbf{x}_0)$ is not periodic. In Chapter 6 we will prove that the limit set $L(\mathbf{x}_0)$ is the entire unit circle $C = \{(x, y) : x^2 + y^2 = 1\}$. Here we offer some numerical evidence of this fact. We have plotted 250 iterations starting from $(1, 0)$. It seems that the orbit is slowly filling up the entire unit circle.

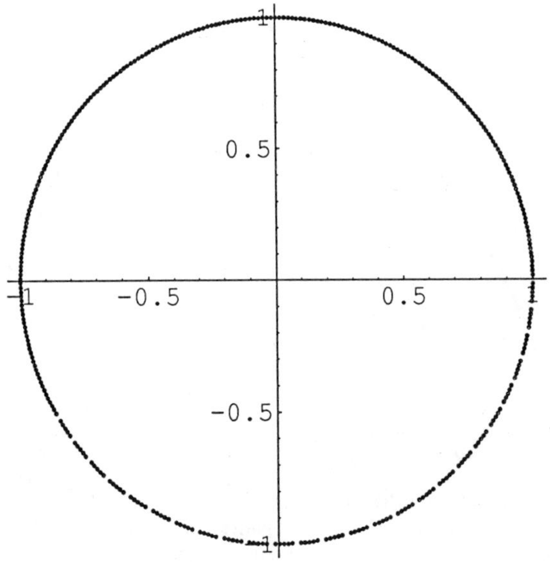

Fig. 1.1.4 For the dynamical system governed by the map $F(x, y) = (x\,\text{Cos}1 - y\,\text{Sin}1, x\,\text{Sin}1 + y\,\text{Cos}1)$ we have plotted the first 250 states of the orbit starting from $(1, 0)$.

The aperiodic character of an orbit is not easy to establish, since it depends on its asymptotic behavior. We shall see later some criteria for making this task easier. The following result ensures the existence of orbits with infinitely many limit points for discrete dynamical systems in the real line.

Theorem of Li and Yorke [Li-Yorke, 1975] *Let* I *be an interval and* $F : I \to I$ *be continuous. Assume that* F *has a periodic orbit of period* 3. *Then* F *has a periodic orbit of every period and there is an infinite set* S *contained in* I *such that every orbit starting from a point of* S *is aperiodic.*

The criterion looks simple, but its application to specific cases may be challenging.

Example 1.1.23 Consider the map $F : [0, 1] \to [0, 1]$ defined by

$$F(x) = \begin{cases} 3x & \text{for } 0 \le x \le \frac{1}{3} \\ \frac{17}{9} - \frac{8}{3}x & \text{for } \frac{1}{3} < x \le \frac{2}{3} \\ \frac{1}{9} & \text{for } \frac{2}{3} < x \le 1 . \end{cases}$$

The map is continuous and

$$F(\tfrac{1}{9}) = \tfrac{1}{3}, \; F(\tfrac{1}{3}) = 1, \; F(1) = \tfrac{1}{9} .$$

Hence the point $1/9$ is periodic of prime period 3. According to the theorem of Li and Yorke there is an infinite set S such that the orbit of every point of S has infinitely many limit points. Determining the elements of S is not a trivial task.

We could define "chaotic" any dynamical system with aperiodic trajectories. We shall see, however, that this definition is not appropriate since it qualifies as chaotic dynamical systems which do not belong to this category. Therefore, the definition presented later (see Chapter 6) is based on requirements which are more stringent than the simple presence of aperiodic orbits.

Section 2. Examples of Discrete Dynamical Systems

In this section we present four examples of discrete dynamical systems taken from the research literature. Our goal is to study the behavior of these systems rather than show how they are derived or to test their appropriateness, objectivity and usefulness. Therefore, we shall keep to a minimum any analysis of the modelling steps used in converting the real dynamical processes into their corresponding mathematical models, and any discussion concerning the interpretation of the conclusions derived from the models in terms of their real counterparts. A detailed study of each system will be provided in the last chapter, using the theoretical results established in the previous chapters. In this section we simply introduce the function governing each system and determine some of the stationary states.

1. One-dimensional example: blood-cell biology

The population of red blood cells in a healthy human individual oscillates, under normal circumstances, within a certain tolerance interval. For men, this interval is $(5.4 \pm 0.8) \times 10^6$ per cubic millimeter, while for women is $(4.8 \pm 0.6) \times 10^6$ per cubic millimeter. In the presence of a disease like anemia, this behavior changes, sometimes dramatically. In modelling this process we are faced with conflicting constraints. On one hand the time appears in the process as a continuous variable, while on the other the number of cells changes only by integer values.

Therefore, in a model based on a continuous approach and which proposes a differential equation on a function $x(t)$ representing cells per unit volume at time t, there is the problem of assessing how well $x(t)$ models its discontinuous real counterpart. Vice versa, in a model where time is discretized, and the behavior of the process is influenced by this discretization procedure, there is the problem of establishing what kind of discretization will preserve the qualitative features of the real process.

The model we present here uses this last approach. Time is divided into discrete generations, numbered $1, 2, ..., n, ...$ and it is assumed that a quantity x_{n+1} representing cells per unit volume at time $n+1$, depends on its state x_n according to the relation

$$(1.2.1) \qquad x_{n+1} = x_n + p_n - d_n .$$

In (1.2.1) p_n and d_n account for the number of cells produced (p_n) and destroyed during the n-th generation. We assume (see [Lasota, 1977]) $d_n = ax_n$, $a \in [0, 1]$, and

$$p_n = b(x_n)^r e^{-sx_n}$$

with b, r, s all positive. Therefore, our one-dimensional discrete dynamical system is

$$(1.2.2) \qquad x_{n+1} = (1 - a)x_n + b(x_n)^r e^{-sx_n}$$

and $F(x, \mathbf{a}) = F(x, (a, b, s, r)) = (1 - a)x + bx^r e^{-sx}$.

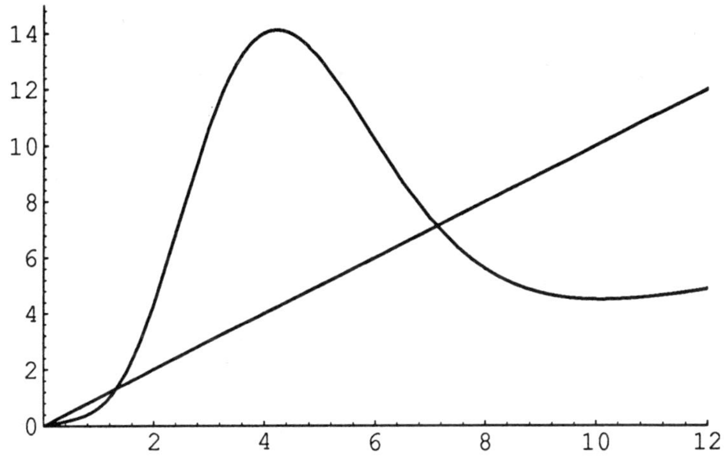

Fig. 1.2.1 Graph of the function F governing the blood-cell population for the following values of the parameters : a = 0.6, b =1, r = 6, and s = 1.45 .

It is easily seen that $F(0, \mathbf{a}) = 0$ for every choice of $\mathbf{a}$. Hence 0 is always a stationary state. In the case a = 1, which means that during the time interval under consideration all red cells which were alive at time n are destroyed, we obtain the simpler model

$$(1.2.3) \qquad x_{n+1} = b(x_n)^r e^{-sx_n} .$$

We shall provide a detailed analysis of (1.2.2) and (1.2.3) in Chapter 7.

2. Two-dimensional example : population biology

During World War I the population of selachians in the north side of the Adriatic Sea rose dramatically, confronting biologists with a seemingly unsolvable puzzle. Umberto D'Ancona, one of them, presented the problem to a mathematics expert, his father-in-law Vito Volterra [Volterra, 1931], who provided a brilliant solution to the puzzle, by building a simple, yet objective and useful model .

He divided the fish population into two categories, predators (the selachians) and prey. Assuming that the number of encounters per unit time between pairs of the two groups is proportional to the product of their sizes and favors the predators' increase, while decreasing the other group, Volterra proposed the following elementary system of differential equation for studying the dynamics of the two populations. The variables $x(t)$, $y(t)$ represent the prey and predators population respectively at time t, and a, b, c, p, q are non-negative parameters:

$$(1.2.4) \qquad \begin{cases} x'(t) = ax(t) - bx(t)y(t) - cx(t) \\ y'(t) = -py(t) + qx(t)y(t) - cy(t)\,. \end{cases}$$

The last terms of the two equations, $-cx(t)$ and $-cy(t)$, account for the fishing activity. Due to war hostilities this activity was significantly reduced from 1915 to 1919, the period characterized by the large increase in the proportion of selachians captured by the few fishing vessels operating in the area.

We can discretize (1.2.4) using the **one-step Euler method.** A time-step $\Delta t = h$ is selected and the two derivatives are replaced by

$$(1.2.5) \qquad \frac{x(t+h) - x(t)}{h} \quad \text{and} \quad \frac{y(t+h) - y(t)}{h}$$

respectively. We arrive at the system

$$(1.2.6) \qquad \begin{cases} x(t+h) = (1 + ah)x(t) - bhx(t)y(t) - chx(t) \\ y(t+h) = (1 - ph)y(t) + qhx(t)y(t) - chy(t)\,. \end{cases}$$

We choose $t = 0$, and an initial state $x(0) = x_0$, $y(0) = y_0$.
Substitution into (1.2.6) gives

$$(1.2.7) \qquad \begin{cases} x_1 = x(h) = (1 + ah - ch)x_0 - bhx_0y_0 \\ y_1 = y(h) = (1 - ph - ch)y_0 + qhx_0y_0\,. \end{cases}$$

We choose $t = h$, and from (1.2.6) and (1.2.7) we obtain

$$(1.2.8) \qquad \begin{cases} x_2 = x(2h) = (1 + ah - ch)x_1 - bhx_1y_1 \\ y_2 = y(2h) = (1 - ph - ch)y_1 + qhx_1y_1\,. \end{cases}$$

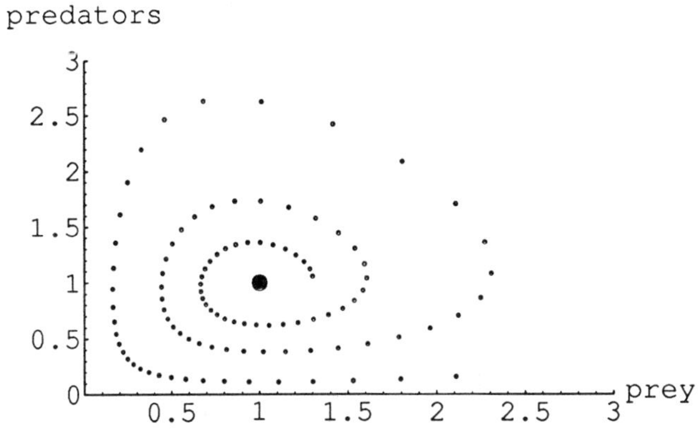

Fig. 1.2.2 Plot of predators versus prey for (1.2.9) with 1+ha = 1.2, bh = 0.2, c = 0, 1-ph = .8, qh = .2. The initial condition is $x_0 = 1.3$, $y_0 = 1$. Noticeable is the "spiraling out" behavior as the number of iterations increases.

In general, after n+1 steps we arrive at

$$(1.2.9) \qquad \begin{cases} x_{n+1} = x(nh) = (1 + ah - ch)x_n - bhx_ny_n \\ y_{n+1} = y(nh) = (1 - ph - ch)y_n + qhx_ny_n . \end{cases}$$

In the discrete dynamical system (1.2.9) we have $\mathbf{x} = (x, y)$, $\mathbf{a} = (a, b, c, h, p, q)$ and

$$F(\mathbf{x}, \mathbf{a}) = ((1+ ha - ch)x - bhxy, (1 - ph - ch)y + qhxy).$$

We see that (0, 0) is always a stationary state of our system. Setting

$$(1.2.10) \qquad x_{n+1} = x_n , \; y_{n+1} = y_n, \qquad x_n \neq 0, \; y_n \neq 0,$$

in (1.2.9) we obtain a second equilibrium point

$$(1.2.11) \qquad x = \frac{p+c}{q} \quad y = \frac{a-c}{b} .$$

The dynamical system (1.2.9) will be analyzed in the last chapter of the book. In the meantime, observe that by increasing the fishing activity (larger c), the stationary state (1.2.11) changes with an increase in x, which represents the population of food fish, and a decrease in y, which represents the population of predators. The result is in agreement with experimental data .

We should not consider system (1.2.9) as a model of the biological process studied by Umberto D'Ancona, particularly if we are interested in the long range evolution of the process. The discretization technique provides only an approximate version of (1.2.4). Given an initial condition and selecting h sufficiently small we can make sure that the trajectory of (1.2.9) remains sufficiently close to the corresponding orbit of (1.2.4) during a selected time interval. However, as the number of iterations increases, the accuracy decreases. The long range behavior of the two systems is different. The orbits of the continuous version are closed, smooth curves around the non-trivial equilibrium point, while the orbits of the discrete system are increasing spirals which eventually exit the first quadrant. This situation gives the warning signal that numerical methods must be applied with great care to make sure that the solutions obtained with them are not artifacts of the numerical scheme.

Other discrete models of predator-prey interaction have been proposed. For example, the following dynamical system in $\mathbf{R}^2$ belongs to this class,

$$(1.2.12) \qquad \begin{cases} x_{n+1} = ax_n(1 - x_n - y_n) \\ y_{n+1} = ay_nx_n . \end{cases}$$

We will analyze (1.2.9) and (1.2.12) in Chapter 7.

3. Three-dimensional example : meteorology

In 1963, the meteorologist E. N. Lorenz [Lorenz, 1963] proposed a simplified model of atmospheric behavior which involved only three differential equations in the three quantities $x(t)$, $y(t)$ and $z(t)$.

To understand the meaning of these variables, let us imagine that in the upper atmosphere there is a layer of air of uniform depth H, where the temperature difference between the upper and lower surface is maintained at a constant value ΔT. In an ideal situation the temperature fall from the upper to the lower surface is linear, i.e. the graph of T versus h, from h_0 to $h_0 + H$, is a straight line with negative slope. In general, there will be a deviation from this linear profile. The variable z is proportional to this distortion. The variable x is proportional to the intensity of the convective motion between the two layers, while the variable y is proportional to the temperature difference between the ascending and descending currents.

With these variables at hand, and after a clever manipulation of a much more complex system involving nonlinear Partial Differential Equations, Lorenz arrived at the simplified model

$$(1.2.13) \qquad \begin{cases} x' = -ax + ay \\ y' = -xz + rx - y \\ z' = xy - bz. \end{cases}$$

Discretization of (1.2.13) using the one-step Euler method gives the dynamical system in $\mathbf{R}^3$

$$(1.2.14) \qquad \begin{cases} x_{n+1} = (1 - ha)x_n + hay_n \\ y_{n+1} = -hx_n z_n + hrx_n + (1 - h)y_n \\ z_{n+1} = hx_n y_n + (1 - bh)z_n \end{cases}$$

with $h = \Delta t$, $\mathbf{x} = (x, y, z)$, $\mathbf{a} = (a, b, r, h)$ and

$$F(\mathbf{x}, \mathbf{a}) = ((1 - ha)x + hay, - hxz + hrx + (1 - h)y, hxy + (1 - bh)z).$$

We see that $(0, 0, 0)$ is always an equilibrium point, regardless of the value of the parameter $\mathbf{a}$. The other equilibrium points are found by setting

$$x_{n+1} = x_n, \; y_{n+1} = y_n, \; z_{n+1} = z_n$$

in (1.2.14) . We obtain from the first equation $x_n = y_n$. Substituting into the second and third equation we arrive, after dropping the subscripts, at the system

$$\begin{cases} xz = rx - x \\ x^2 = bz. \end{cases}$$

The first equation gives $z = r - 1$, since $x = 0$ implies $y = z = 0$. Thus $x^2 = b(r - 1)$, which requires $r > 1$ ($r = 1$ gives again $x = y = z = 0$) . We find two more stationary states

$$E_1 = (\sqrt{b(r - 1)}, \sqrt{b(r - 1)}, r - 1), \; E_2 = (- \sqrt{b(r - 1)}, - \sqrt{b(r - 1)}, r - 1).$$

The comments made about the discretization of the continuous model proposed by Volterra to solve the problem of D'Ancona, should be repeated here for the discretization of the model of Lorenz. Here however, the qualitative behavior of the model and of its discrete approximation are not as different as in the previous case.

We shall analyze this system in the last chapter.

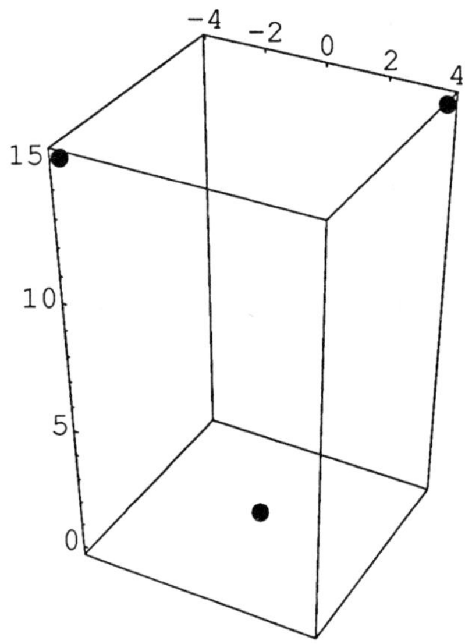

Fig. 1.2.3 The three stationary states of the Lorenz system for $a = 8$, $b = 1$ and $r = 16$.

4. **Multi-dimensional example : neural networks**

A neural network is a crude but powerful model of the human brain, a kind of electronic device that attempts to mimic the way neurons of the brain work together. The potential of neural networks is still largely unknown, but the expectations are very high, based on the success of the applications in place so far.

Chase Manhattan Bank uses a neural net system to detect credit card frauds, while at Security Pacific a neural system analyzes loan risks. Our fastest available computers take several years of CPU time to solve problems that neural nets can answer in a few seconds. There is a price to pay: the computer solution is exact, while the neural network solution has a 5-10% margin of error. However, we do not have many choices, since in many circumstances we need the solution within a few seconds after the problem has been posed.

Our brain has billions of neurons, but real neural nets are far from this level. If we imagine a net with N neurons, then the connections among them can be arranged in a two-dimensional square array, called a **connectivity matrix** T, with N^2 entries. The entry t_{ij} located in row i and column j of T, represents the action of neuron j on neuron i. This action can be **excitatory** ($t_{ij} > 0$) or **inhibitory** ($t_{ij} < 0$). When neuron j and i are not directly connected we have $t_{ij} = 0$.

A function f, called "neuron response function", controls the action of each neuron, and works like a switch. As long as the energy of neuron j remains below a threshold level, the state of the neuron is purely passive. The neuron can be influenced by other neurons, but cannot have any effect on their state. When the threshold level is reached the function f triggers its response.

The dynamical evolution of a neural net after receiving an initial input $\mathbf{x}_I$, which usually remains active during the entire process, can be modelled by the system of equations (see [Hopfield, 1982])

$$x_{k,n+1} = x_{k,n} - c_k x_{k,n} + t_{k,1} f(x_{1,n}) + \dots + t_{k,N} f(x_{N,n}) + x_{I,k} \tag{1.2.15}$$

for k = 1, 2, ..., N. Equation (1.2.15) represents the energy level of neuron k at the n+1 stage of the evolution of the system. The quantity c_k, $0 < c_k < 1$, accounts for a certain leakage of neuron's k energy; the quantities $t_{k,j}$ for j = 1, 2, ..., N are the connections between all other neurons and neuron k, and $f(x_{j,n})$, j = 1, 2, ..., N controls neuron's j response at the n-th state. We assume here that all neuron response functions are the same. The term $x_{I,k}$ accounts for the effect of the input $\mathbf{x}_I$ on neuron k. As we said before, we want the input to remain active throughout the entire process. With a net of 100 neurons we have a system of 100 equations!

In the last chapter of this book we shall analyze certain features of (1.2.15). In the meantime we may think that the net provides an answer to our problem when the dynamical process defined by (1.2.15) reaches a stationary state.

The entries of the connectivity matrix are obviously a crucial component of the process. Their optimal choice can be achieved in two ways. The first is to select certain learning rules and then let the net teach itself how to perform correctly the assigned tasks by adjusting the connections among the various neurons.

The second is to use an external optimization procedure based on an energy function whose variables are the entries of T. The connections are then determined by requiring that the correct execution of each task corresponds to a minimum of the energy .

Problems

Section 1.

1.1.1 Let $x_{n+1} = ax_{n-1}(1 - x_{n-2})$. Replace this one-dimensional system with an equivalent three-dimensional system with no delay.

1.1.2 Find the stationary states and the periodic orbits of period 2 of $F(x, a) = -ax^2 + 1$.

1.1.3 Let $F(x, a) = -ax^2 + 1$. What is the degree in x of the polynomial $F^3(x, a)$? And what about the degree in a ? Generalize to F^n, the n-th iterate of F.

1.1.4 Let $F(x, a) = -ax^2 + 1$. Show that for $0 < a < 2$, F maps the interval $[-1, 1]$ into itself, i.e. for every $x \in [-1, 1]$ we have $F(x, a) \in [-1, 1]$.

1.1.5 Let $F(x, a) = -ax^2 +1$ and let $x_0 = 1$. Find x_1 and x_2. Determine $\frac{d}{dx}F^3(x, a)$ at the point $x = 1$. Recall that by $F^3(x, a)$ we mean the third iterate of $F(. , a)$.

1.1.6 Let $p = aq(1 - q)$, $q = ap(1 - p)$. Verify that $p + q = (a + 1)/a$, $pq = (a + 1)/a^2$. Use these equalities to find the periodic orbits of period 2 of $F(x, a) = ax(1 - x)$.

1.1.7 Consider the map of Example 1.1.23. Show that the limit set of every orbit starting from a point x, $\frac{2}{3} \le x \le 1$, is $L(x) = \{\frac{1}{9}, \frac{1}{3}, 1\}$.

1.1.8 Let $F(x) = x(1 - x)$. Prove that for every $x \in [0, 1]$, we have $L(x) = \{0\}$.

1.1.9 In Example 1.1.2 assume first that there are no predators ($y = 0$) and verify that given an initial condition x_0, we have $x_n = a^n x_0$. Is this unrestricted growth reasonable ?

1.1.10 In Example 1.1.2 assume that at certain point in time prey become extinct. Denote by y_0 the predators' population at that time and verify that the species will become extinct .

1.1.11 In the search for stationary states of the predator-prey model, verify that $x = 0$ if and only if $y = 0$ (recall that $a > 1$ and $c \in (0, 1)$).

1.1.12 Assume that a dynamical system governed by a map $F(., \mathbf{a})$ has exactly two periodic points of prime period 2, $\mathbf{x}_1$ and $\mathbf{x}_2$, for a certain value of the parameter **a**. Prove that $F(\mathbf{x}_1, \mathbf{a}) = \mathbf{x}_2$ and $F(\mathbf{x}_2, \mathbf{a}) = \mathbf{x}_1$.

1.1.13 Let $F(x, y) = (-y, x)$. Verify that the origin is a stationary state and that every other point is periodic of prime period 4.

1.1.14 Verify that each periodic point of period p is a stationary state of F^{np} for every integer $n = 1, 2, \ldots$.

1.1.15 Let $F(x) = \frac{3}{2} x (1 - x)$. Verify that $x_s = \frac{1}{3}$ is a stationary point. Prove that for every $x \neq 0, 1$ we have $x_n \to \frac{1}{3}$ as $n \to \infty$.

Section 2.

1.2.1 The map $F(x, y) = (1 + y - ax^2, bx)$ was studied by M. Hénon [Hénon, 1976]. Find the stationary states of the Hénon map.

1.2.2 Write the complete set of equations (1.2.15) when $N = 4$.

1.2.3 Assume that the response function of neuron i has the form $f_i(x) = a_i x + b_i$, $i = 1, 2, \ldots, N$. Rewrite system (1.2.15) for $N = 3$.

1.2.4 Choose $a = 1.2$, $b = 0.2$, $c = 0$, $p = .6$, $q = 0.2$, $h = 1$ in the discretized model of competition among two species. Select an initial state (x_0, y_0), with $x_0 > 0$ and $y_0 > 0$. Investigate numerically the behavior of the point

$$(z_n, w_n) = \frac{1}{n} (x_0 + x_1 + \ldots + x_{n-1},\ y_0 + y_1 + \ldots + y_{n-1})$$

as n grows. Does (z_n, w_n) approach the non-trivial equilibrium point ?

1.2.5 Let $F(x, y) = (x + |x - y| - 2, y - x + 1)$. Find the equilibrium points of the dynamical system defined by F.

1.2.6 Verify that the Hénon map is one-to-one, provided that $b \neq 0$.

1.2.7 Find the stationary states of the system

$$\begin{cases} x_{n+1} = ax_n(1 - x_n - y_n) \\ y_{n+1} = by_n x_n . \end{cases}$$

1.2.8 Let $x_{n+1} = (1 - a)x_n + 4x_n^6 \exp(-2x_n)$. Verify that for every $a \in (0, 1)$ the system has exactly three equilibrium points.

CHAPTER 2

ONE-DIMENSIONAL DYNAMICAL SYSTEMS

SUMMARY

In this chapter we present an extensive analysis of discrete one-dimensional systems, starting with some ideas on conjugacy and on linear or affine systems (section 1). We then turn our attention to non-linear systems and we provide, in section 2, sufficient criteria for the stability of a stationary state and of a periodic orbit. Section 3 is reserved to a further investigation on the consequences of conjugacy between two systems. In the last section of the chapter we study bifurcation points and bifurcation diagrams.

Section 1. Preliminaries on One-Dimensional Systems

A discrete dynamical system is said to be one-dimensional if it is of the form

$$(2.1.1) \qquad x_{n+1} = F(x_n, \mathbf{a})$$

where $F(., \mathbf{a}) : I_a \to \mathbf{R}$ usually maps an interval I_a into itself. The interval I_a may not depend on the parameter $\mathbf{a}$. In this case we write $F : I \to I$. The parameter $\mathbf{a}$ is frequently a number (a scalar), but this restriction should not be imposed in general. Unless otherwise stated we assume that F and F_x are continuous.

Example 2.1.1 Let $F(x, a) = ax(1 - x)$, $x \in [0, 1]$. Then $F_x(x, a) = a - 2ax$ and F, F_x are continuous. For every $a \in [0, 4]$ the dynamical system

$$(2.1.2) \qquad x_{n+1} = ax_n(1 - x_n)$$

represents the evolution of a process in the interval [0, 1]. We could identify it with a simplified model of the dynamics of a population which lives in a limited environment. The growth of the population is controlled by the parameter a, which may be considered as the vital coefficient of the population.

One could think to obtain a more realistic model by considering the dynamical system

$$(2.1.3) \qquad x_{n+1} = x_n(a - bx_n) \ .$$

However, the improvement is only apparent since a suitable change of variable reduces the study of (2.1.3) to the study of (2.1.2) (see Problem 2.1.1).

1. The cobweb method

Assume that in (2.1.1) the parameter $\mathbf{a}$ is fixed. We can perform an analysis of the dynamical system governed by $F(., \mathbf{a})$ using the graph of the function F.

For simplicity, denote $F(., \mathbf{a})$ by F. The fixed points of F are the intersections of its graph with the line $y = x$. The trajectory starting from x_0 can be visualized in the plane by drawing a vertical segment from the point $(x_0, 0)$ to $(x_0, F(x_0)) = (x_0, x_1)$ on the graph of F; then horizontally to the line $y = x$ to meet the point (x_1, x_1), and then again vertically to $(x_1, F(x_1)) = (x_1, x_2)$ on the graph of F. Continue this process until sufficient information is obtained on the behavior of the trajectory starting from x_0. The picture which provides this information is called a **cobweb**.

For example, when the orbit $O(x_0)$ converges to the stationary state x_s, the cobweb starting from $(x_0, 0)$ will be a sequence of vertical and horizontal segments spiraling towards the point (x_s, x_s). In Fig. 2.1.1 the situation is illustrated by a trajectory of the logistic map $F(x, a) = ax(1 - x)$ for $a = 2.9$. The stationary state is $x_s = \frac{19}{29}$.

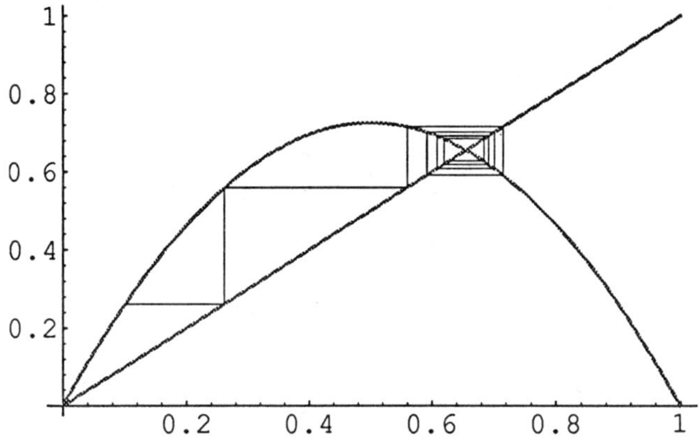

Fig. 2.1.1 The cobweb methods shows how the orbit starting from $x_0 = .1$ converges to an equilibrium point in the dynamical system governed by $F(x) = 2.9x(1 - x)$.

Similarly, when the orbit $O(y_0)$ converges to the periodic orbit $O(x_0, p)$ the cobweb starting from $(y_0, 0)$ will approach the closed cycle

$$(x_0, x_0), (x_0, x_1), \ldots, (x_{p-1}, x_{p-1}), (x_{p-1}, x_0), (x_0, x_0).$$

In Fig. 2.1.2 we illustrate this situation by selecting the logistic map $F(x, a) = ax(1 - x)$ with $a = 3.4$. In this case, as we shall see later, there is an attractive periodic orbit of prime period 2. The cobweb method illustrates how the orbit starting from $x_0 = .1$ converges to the periodic orbit.

Finally, when the orbit starting from x_0 is aperiodic, the behavior will be graphically illustrated by the cobweb method. The sequence of vertical and horizontal segments will not display any recognizable pattern. In Fig. 2.1.3 we have selected $a = 4$ in the logistic map. The initial condition is once again $x_0 = .1$.

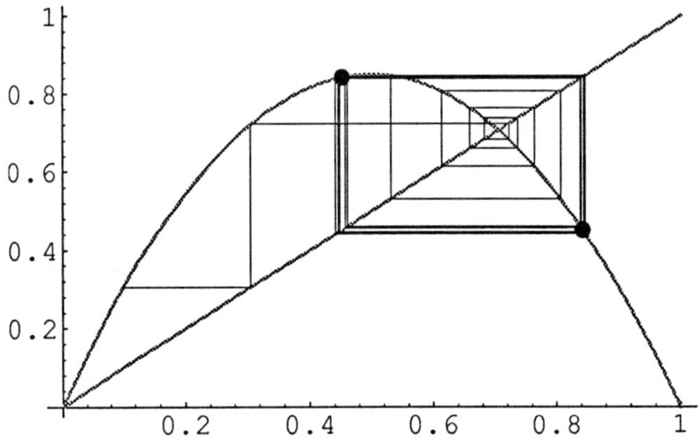

Fig. 2.1.2 The orbit starting at $x_0 = .1$ passes very close to the stationary state $\frac{12}{17}$ of the dynamical system governed by $F(x) = 3.4x(1 - x)$, before approaching the attractive periodic orbit of prime period 2.

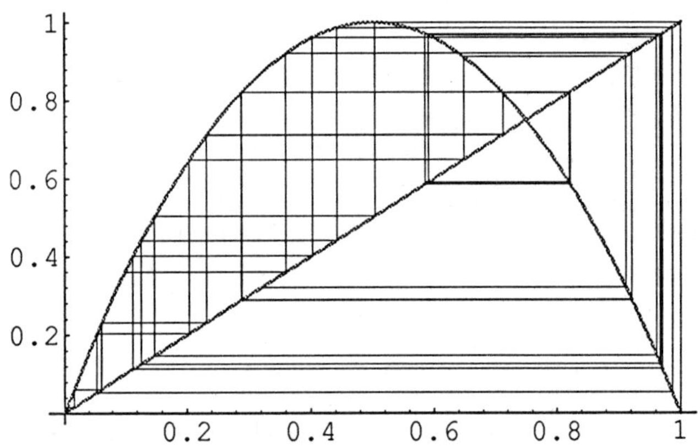

Fig. 2.1.3 Starting at $x_0 = .1$ we follow the orbit of the dynamical system $F(x, 4) = K(x) = 4x(1 - x)$. No pattern seems to emerge. The orbit is apparently aperiodic.

2. Conjugacy

We have mentioned previously that a change of variable may sometimes reduce a system to a simpler one. This idea is made more precise with the concept of conjugate maps.

Definition of conjugate maps

Let I and J be two intervals and $h : I \to J$ be continuous, one-to-one and onto. We say that h is a **conjugacy** between $F : I \to I$ and $G : J \to J$ if

(2.1.4) $$h(F(x)) = G(h(x))$$

for all $x \in I$. The maps F and G are said to be **conjugate** (by h). Since h is invertible, (2.1.4) implies

(2.1.5) $$F = h^{-1}Gh\ ,$$

and in general,

(2.1.6) $$F^n = h^{-1}G^nh \quad \text{or} \quad hF^n = G^nh.$$

The study of the dynamical system generated by the map F, i.e. the analysis of the trajectories of

(2.1.7) $$x_n = F(x_{n-1}) = F^n(x_0), \qquad\qquad x_0 \in I\ ,$$

can be performed by investigating the orbits of the system generated by G. In fact, from (2.1.6) we obtain, for all $x_0 \in I$,

(2.1.8) $$x_n = F^n(x_0) = h^{-1}(G^n(h(x_0))).$$

In particular, the h^{-1}-image of a fixed point y_s of G, $x_s = h^{-1}(y_s)$, is a fixed point of F since

$$F(x_s) = h^{-1}G(h(x_s)) = h^{-1}G(y_s) = h^{-1}(y_s) = x_s.$$

Similarly, the h^{-1}-image of a periodic orbit $\{y_0, ..., y_{p-1}\}$ of prime period p of G, is the periodic orbit $\{x_0 = h^{-1}(y_0), ..., x_{p-1} = h^{-1}(y_{p-1})\}$ of prime period p of F. In fact

$$F^p(x_0) = h^{-1}(G^p(h(x_0))) = h^{-1}(G^p(y_0)) = h^{-1}(y_0) = x_0\ ,$$

and for every $q < p$,

$$F^q(x_0) = h^{-1}(G^q(h(x_0))) = h^{-1}G^q(y_0) \neq h^{-1}(y_0) = x_0\ .$$

At the end of this chapter we shall analyze the implications of conjugacy with respect to stability and instability of stationary states and periodic orbits. In the meantime we illustrate the usefulness of this concept with some examples.

Example 2.1.2 We want to study the dynamical system generated in the real line by the map $F(x) = ax + b$. The numbers a, b are fixed.

Assume that $a \neq 1$ and set $y = h(x) = x - \frac{b}{1-a}$. Then $F(x) = h^{-1}(G(h(x)))$, where $G(y) = ay$. We shall see shortly that the study of the trajectories of the dynamical system generated by G is very simple.

The case $a = 1$ can be analyzed directly since the map $F(x) = x + b$ represents a translation. Hence, for $b > 0$ every trajectory diverges to $+\infty$, while for $b < 0$ every trajectory diverges to $-\infty$. For $b = 0$ every point is a stationary state.

Example 2.1.3 Let $I = [-1, 1]$ and let $F, G : I \to I$ be defined by

$$F(x) = -2x^2 + 1, \; G(x) = -2|x| + 1.$$

The maps F and G are conjugate by $h : I \to I$, $h(x) = \frac{2}{\pi}\text{ArcSin}x$, since $h(F(x)) = G(h(x))$ for every $x \in [-1, 1]$. A graphical plot of h, hF and Gh is the easiest way to verify this result, although everything can be established theoretically .

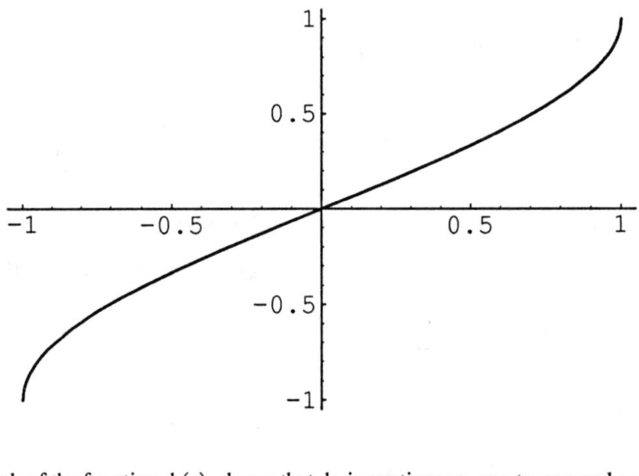

Fig. 2.1.4 The graph of the function $h(x)$ shows that h is continuous, one-to-one and onto.

The only fixed points of the map G are $y = -1$ and $y = \frac{1}{3}$. Thus $h^{-1}(-1) = -1$ and $h^{-1}(\frac{1}{3}) = \frac{1}{2}$ are the only fixed points of F. They can be directly obtained from the quadratic equation $-2x^2 + 1 = x$.

The periodic orbit of prime period 2 of G, $\{y_0 = -\frac{1}{5}, y_1 = \frac{3}{5}\}$ is mapped into the periodic orbit of prime period 2 of F : $\{x_0 = -\mathrm{Sin}(\pi/10), x_1 = \mathrm{Sin}(3\pi/10)\}$.
A similar approach can be used to find periodic orbits of prime period 3, 4, ... of F.

In Example 2.1.3 the study of the dynamical system governed by G is certainly easier than the analysis of the one generated by F. The conjugacy allows us to transfer the results found for G into corresponding results for F.

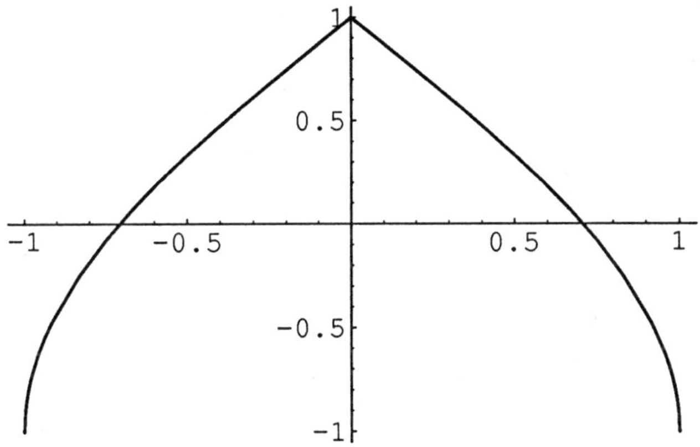

Fig. 2.1.5 The graph of h(F(x)), x ∈ [-1, 1].

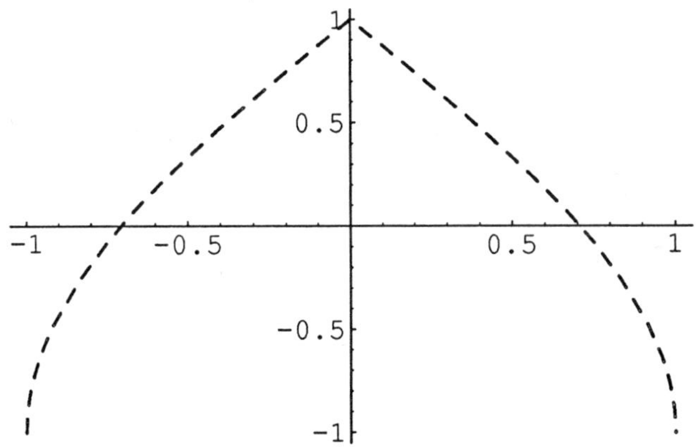

Fig. 2.1.6 The graph of G(h(x)), x ∈ [-1 , 1]. The graphs of Fig. 2.1.5 and Fig. 2.1.6 constitute enough evidence that the two compositions coincide.

3. Linear and affine systems

Assume that in (2.1.1) the function F has the form F(x, **a**) = ax + b. Then the dynamical system is said to be **linear** (b = 0) or **affine** (b ≠ 0). In Example 2.1.2, we have seen that the study of the system

(2.1.9) $$x_{n+1} = ax_n + b, \quad a \neq 1,$$

can be performed by analyzing the system

(2.1.10) $$x_{n+1} = ax_n .$$

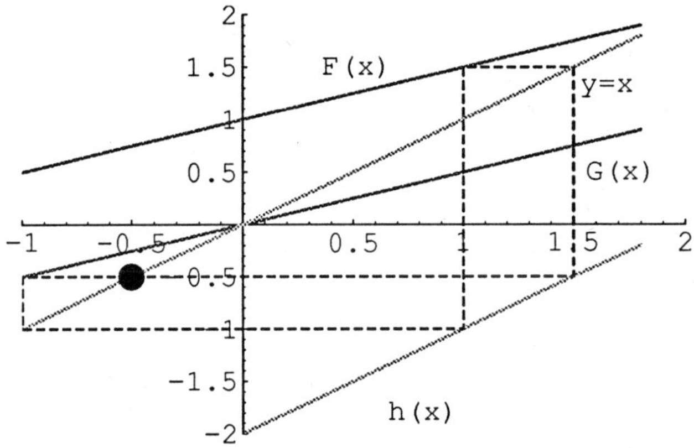

Fig. 2.1.7 The dynamical system governed by the affine function F(x) = .5x + 1 is conjugate to the one governed by the linear function G(x) = .5x. The cobweb diagram shows that h(F(1)) = G(h(1)).

The orbits of (2.1.10) are of the form $\{a^n x_0 : n = 1, 2, ..., x_0 \in \mathbf{R}\}$. Therefore, if $|a| < 1$ every orbit converges to the origin. When $|a| > 1$, every non stationary orbit diverges to infinity. For $a = 1$ every orbit is stationary, while for $a = -1$ every non stationary orbit is periodic of period 2.

These results for (2.1.10) provide corresponding results for (2.1.9). The stationary state of (2.1.9) is $x_s = \dfrac{b}{1-a}$. For $|a| < 1$ every orbit of (2.1.9) converges to x_s, while for $|a| > 1$ every non stationary orbit diverges to infinity. For $a = -1$ every non stationary orbit is periodic of prime period 2.

Section 2. Nonlinear One-Dimensional Systems

We now study nonlinear one-dimensional dynamical systems. We first analyze local stability and instability of equilibrium points and periodic orbits.

1. Stationary states and periodic orbits

The equilibrium states of the dynamical system (2.1.1) are the solutions of the equation

(2.2.1) $$x = F(x, \mathbf{a}) .$$

Example 2.2.1 Let $F(x, a, b) = ax^3 - bx$. The stationary states are obtained from the equation $x = ax^3 - bx$, whose solutions are $x = 0$, and $x = \pm\sqrt{\frac{1+b}{a}}$.

Similarly, the periodic points of prime period p are the solutions of the equation

(2.2.2) $$x = F^p(x, \mathbf{a}),$$

which are not solutions of the equation

(2.2.3) $$x = F^m(x, \mathbf{a})$$

for all $m < p$.

Example 2.2.2 The periodic solutions of prime period 2 of the dynamical system generated by $F(x) = 4x(1 - x)$ are the solutions of the equation

$$x = 4[4x(1 - x)\ (1 - 4x(1 - x))],$$

which are not solutions of the equation

$$x = 4x(1 - x),$$

i.e. $x \neq 0$ and $x \neq 3/4$. Deleting the solution $x = 0$ we obtain

$$1 = 16(1 - x)(1 - 4x + 4x^2) \text{ or } 64x^3 - 128x^2 + 80x - 15 = 0.$$

Dividing by $x - \frac{3}{4}$ we obtain $16x^2 - 20x + 5 = 0$, whose solutions are

$$x_0 = \frac{5 - \sqrt{5}}{8}, \quad x_1 = \frac{5 + \sqrt{5}}{8} .$$

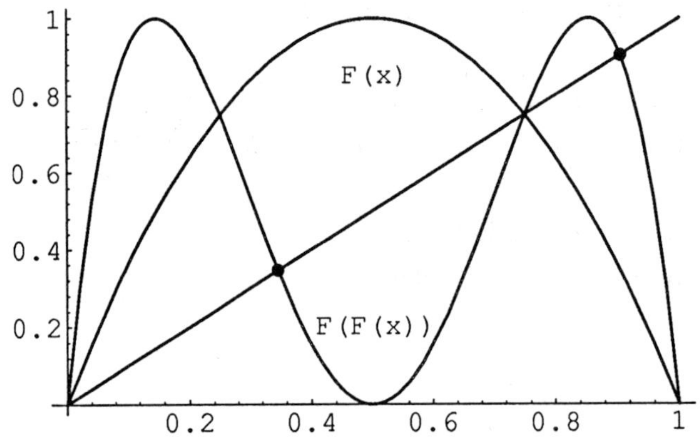

Fig. 2.2.1 Graph of $F(x) = 4x(1 - x)$ and of its second iterate $F(F(x))$. There are two periodic points of period two, as detected by the analytic study of the example.

These coincide with the solutions given in Example 1.1.13 for $a = 4$. The points x_0, x_1 are the only periodic points of prime period 2. Consequently $F(x_0) = x_1$ and $F(x_1) = x_0$.

2. **Stability and instability of stationary states and periodic orbits**

Definition of sink

Intuitively we imagine a sink as a stationary state x_s (or a periodic orbit $O(x_0, p)$) which attracts the orbit of every point x_0 close by. We expect $O(x_0)$ to remain close to x_s (x_s is stable), and in addition the sequence $\{|x_n - x_s| : n = 0, 1, ...\}$ to be decreasing and convergent to 0. The formal definition we propose insures this property (see Remark 2.2.1).

Let $F : I \to I$ be continuous on I and differentiable except possibly at finitely many points. We say that an equilibrium point $x_s \in I$ is a **sink** or an **attractor** if there exists $r \in [0, 1)$ and an open bounded interval J containing x_s, such that the inequality $|F_x(x)| \leq r$ holds for all $x \in J \cap I$ where F is differentiable.

We say that a periodic orbit of prime period p, $O(x_0, p)$, is a **sink** or an **attractor** if each of its states is a sink for the dynamical system generated by F^p.

Here and in the sequel we use the following results from Calculus.

Lemma 2.2.1 *Let* I *be an interval and* $g : I \to \mathbf{R}$ *be continuous at a point* $x_0 \in I$. *Assume that* $|g(x_0)| < 1$. *Let* r *be a real number such that* $|g(x_0)| < r < 1$.
Then there exists an open and bounded interval J, *containing* x_0, *such that*

$$(2.2.4) \qquad |g(x)| \le r \quad \textit{for all} \quad x \in J \cap I.$$

The Lemma is still true if < and ≤ are replaced by > and ≥ respectively. Moreover for both versions to be true it is enough that

$$\lim_{x \to x_0} g(x) = L \ , \ |L| < 1 \ \text{ or } |L| > 1.$$

Lemma 2.2.2 (Mean Value Theorem) *Let* $g : [a, b] \to \mathbf{R}$ *be continuous on* [a, b] *and differentiable on* (a, b). *Then there exists* $c \in (a, b)$ *such that*

$$(2.2.5) \qquad g(b) - g(a) = g_x(c)(b - a).$$

Lemma 2.2.3 (Mean Value Inequality) *Let* I *be an interval and let* $g : I \to \mathbf{R}$ *be continuous on* I. *Assume that* g *is differentiable at all but finitely many points of* I. *Then for every pair of points* $x < y \in I$ *there exists a point* $c \in (x, y)$, *such that*

$$(2.2.6) \qquad |g(x) - g(y)| \le |g_x(c)|\, |x - y| .$$

We can now establish a condition for x_s to be a sink.

Theorem 2.2.1 *Let* $F : I \to I$ *be continuous on* I *and let* x_s *be a fixed point of* F. *Assume that there is an open bounded interval* J *containing* x_s *such that* F *is differentiable at every point of* $J \cap I$ *with a continuous derivative at* x_s.
Then $|F_x(x_s)| < 1$ *implies that* x_s *is a sink.*

Proof. Choose a number $r \in (|F_x(x_s)|, 1)$. By Lemma 2.2.1, one can find an open and bounded interval J containing x_s such that $|F_x(x)| \le r$ for all $x \in J \cap I$.

Remark 2.2.1 In the case when x_s is a sink we can estimate how fast a sequence of iterates starting from a point $x_0 \in J \cap I$ converges to x_s. In fact, using the Mean Value Inequality, we obtain ($r \in [0, 1)$ comes from the definition of sink)

$$|x_1 - x_s| = |F(x_0) - x_s| = |F(x_0) - F(x_s)| \le r|x_0 - x_s| ,$$

$$|x_2 - x_s| = |F(x_1) - x_s| = |F(x_1) - F(x_s)| \le r|x_1 - x_s| \le r^2|x_0 - x_s| ,$$

and in general,

$$|x_n - x_s| = |F(x_{n-1}) - x_s| = |F(x_{n-1}) - F(x_s)| \leq r|x_{n-1} - x_s| \leq r^n|x_0 - x_s|.$$

From $|x_0 - x_s| \leq |x_0 - x_1| + |x_1 - x_s|$ we derive $|x_0 - x_s|(1 - r) \leq |x_1 - x_0|$. Therefore

$$(2.2.7) \qquad |x_n - x_s| \leq \frac{r^n}{1 - r}|x_1 - x_0|.$$

This gives an estimate of how close x_n is to x_s without an explicit knowledge of x_s.

Example 2.2.3 Let $F(x, a, b) = ax^3 - bx$. Regardless of the choice of the parameters a and b the point $x_s = 0$ is a stationary state of the system governed by F. Since

$$F_x(x, a, b) = 3ax^2 - b,$$

we have $F_x(0, a, b) = - b$. Thus, for $|b| < 1$, $x_s = 0$ is a sink.

For periodic orbits of prime period p the condition for being a sink is established using F^p in place of F. Recall that each point of the periodic orbit

$$x_0, x_1 = F(x_0), \ldots, x_{p-1} = F(x_{p-2}), x_p = F(x_{p-1}) = x_0, \ldots$$

is a fixed point of F^p. Using the chain rule and the commutativity of the product among real numbers, one obtains

$$\frac{d}{dx} F^p(x_i) = F_x(x_0)F_x(x_1)\ldots F_x(x_{p-1}), \quad 0 \leq i \leq p-1,$$

i.e. the product of the derivatives of F at all points of the periodic orbit. Consequently, we have the fundamental equality

$$(2.2.8) \qquad \frac{d}{dx} F^p(x_i) = \frac{d}{dx} F^p(x_j), \qquad 0 \leq i, j \leq p-1.$$

For example, for p = 2

$$\frac{d}{dx} F^2(x_0) = F_x(F(x_0))\, F_x(x_0) = F_x(x_1)F_x(x_0) = F_x(x_0)F_x(x_1) = \frac{d}{dx} F^2(x_1).$$

As a consequence of these remarks and of Theorem 2.2.3 we obtain the following result.

Theorem 2.2.2 *Let* $F : I \to I$ *be continuous on* I *and let*

$$x_0, x_1 = F(x_0), ..., x_p = F(x_{p-1}) = x_0, ...$$

be a periodic orbit of prime period p. *Assume that there are open intervals* J_i, $x_i \in J_i$, *such that* F *is differentiable on* $J_i \cap I$ *with a continuous derivative at* x_i, $i = 0, ..., p-1$. *Then the inequality*

$$(2.2.9) \qquad \left|\frac{d}{dx} F^p(x_0)\right| < 1$$

implies that the orbit is an attractor.

Proof. By Theorem 2.2.1 and equality (2.2.8) each point x_i is a sink for the dynamical system defined by $F^p : I \to I$. Hence the periodic orbit is an attractor.

Example 2.2.4 Let $F(x) = 3.2\, x(1 - x)$. Then

$$x_0 = \frac{2.1 + \sqrt{.21}}{3.2}, \quad x_1 = \frac{2.1 - \sqrt{.21}}{3.2}$$

is a periodic orbit of period 2 . From $F_x(x) = 3.2 - 6.4x$, we derive

$$F_x(x_0) = -1 - 2\sqrt{.21} \text{ and } F_x(x_1) = -1 + 2\sqrt{.21}\ .$$

Hence $F^2{}_x(x_0) = (-1 - 2\sqrt{.21})(-1 + 2\sqrt{.21}) = 1 - .84 = .16$. The periodic orbit $\{x_0, x_1\}$ is an attractor.

Definition of weak attractor

Sometimes the derivative of the function governing a dynamical system satisfies a weaker inequality in an open and bounded interval J containing an equilibrium point x_s, in the sense that $|F_x(x)| < 1$ for all $x \in J \cap I$, except at $x = x_s$ where the derivative may not exist or $|F_x(x_s)| = 1$. In this case we say that x_s is a **weak attractor**.

A similar definition can be given for a periodic orbit.

Example 2.2.5 Let $F : \mathbf{R} \to \mathbf{R}$ be defined by $F(x) = \text{Sin}\, x$. The only equilibrium point of the dynamical system governed by F is $x_s = 0$. The derivative of Sin x is Cos x,

$\text{Cos } 0 = 1$ and $0 < \text{Cos } x < 1$ for all $x \in (-1, 1)$ except $x = 0$. Thus the equilibrium point $x_s = 0$ is a weak attractor.

The orbits $O(x_0)$ of points close to a weak attractor x_s converge to x_s and the distance from x_n to x_s decreases as n increases. In other words

$$|x_n - x_s| < |x_{n-1} - x_s| \quad \text{and} \quad x_n \to x_s \text{ as } n \to \infty.$$

The first result follows from the Mean Value Inequality and the second from a theorem which will be proved in section 3.

Stability

In the first chapter we have introduced the notion of stability for a stationary state x_s and for a periodic orbit of prime period p, $O(x_0, p)$. We now provide a sufficient condition which insures such property.

Theorem 2.2.3 *Let* $F : I \to I$ *be continuous on* I. *Let* x_s *be a stationary state and let* J *be a bounded open interval containing* x_s, *such that* F *is differentiable at every point* x *of* $J \cap I$ *except possibly at* x_s *and* $|F_x(x)| \leq 1$. *Then* x_s *is stable.*

Proof. Let $x_0 \in J \cap I$. Then, by the the Mean Value Inequality,

$$|x_1 - x_s| = |F(x_0) - F(x_s)| \leq | x_0 - x_s| \ ,$$

and in general,

$$|x_{n+1} - x_s| = |F(x_n) - F(x_s)| \leq | x_n - x_s| \leq \ldots \leq | x_0 - x_s| \ .$$

Thus, the stationary state x_s is stable.

Example 2.2.6 Let

$$F(x) = \begin{cases} x & \text{for } x \leq 0 \\ x - x^3 & \text{for } x > 0 \, . \end{cases}$$

The origin is a stable equilibrium.

Theorem 2.2.3 can be extended to periodic orbits of prime period p by requiring that for every state x_i of the orbit, $i = 0, 1, \ldots, p-1$, there is an open interval J_i such that

F is differentiable at every point x of $J_i \cap I$ except possibly at x_i and $|F_x(x)| \le 1$.

Definition of source

We expect a source to be an unstable equilibrium x_s (or an unstable periodic orbit $O(x_0, p)$) and to have the additional property that all orbits starting at points close to the source move away from it.
The formal definition insures this property (see Remark 2.2.2)

Let $F : I \to I$ be continuous on I and differentiable except possibly at finitely many points. We say that an equilibrium point $x_s \in I$ is a **source** or a **repeller** if there exists $r > 1$ and an open bounded interval J containing x_s such that for all $x \in J \cap I$ where F is differentiable, we have the inequality $|F_x(x)| \ge r$.

We say that a periodic orbit of prime period p, $O(x_0, p)$, is a **source** or a **repeller** if each of its states is a source for the dynamical system generated by F^p.

Theorem 2.2.4 *Let* $F : I \to I$ *be continuous on* I *and let* x_s *be a fixed point of* F. *Assume that there is an open, bounded interval* J *containing* x_s *such that* F *is differentiable at every point of* $J \cap I$ *with a continuous derivative at* x_s.
Then, $|F_x(x_s)| > 1$ *implies that* x_s *is a source.*

Proof. Choose a number $r \in (1, |F_x(x_s)|)$. By Lemma 2.2.1 one can find an open and bounded interval J containing x_s such that $|F_x(x)| \ge r$ for all $x \in I \cap J$.

Remark 2.2.2 In the case when x_s is a source we can estimate how fast an orbit starting at $x_0 \in J \cap I$ separates from x_s as long as it remains in $J \cap I$. In fact, assume that x_0, ..., x_{n-1} are in $J \cap I$ and F is differentiable on $J \cap I$ except possibly at x_s. Using the Mean Value Theorem we obtain,

$$|x_{i+1} - x_s| = |F_x(c_i)|\, |x_i - x_s| \ge r\, |x_i - x_s|, \qquad i = 0, 1, \ldots, n-1.$$

Thus

$$|x_n - x_s| \ge r\, |x_{n-1} - x_s| \ge r^2\, |x_{n-2} - x_s| \ge \ldots \ge r^n\, |x_0 - x_s|.$$

Example 2.2.7 Let $F(x, a, b) = ax^3 - bx$. Regardless of the choice of the parameters a and b, the point $x_s = 0$ is a stationary state of the dynamical system governed by F.

Since $F_x(x, a, b) = 3ax^2 - b$ we have $F_x(0, a, b) = - b$. The stationary state $x_s = 0$ is a source if $|b| > 1$.

For periodic orbits of prime period p, the condition for being a source is established using F^p in place of F.

Theorem 2.2.5 *Let* $F : I \to I$ *be continuous on* I *and let* $x_0, x_1 = F(x_0), x_2 = F(x_1), \ldots, x_p = F(x_{p-1}) = x_0, \ldots$ *be a periodic orbit of prime period* p. *Assume that there are open intervals* J_i, $x_i \in J_i$, *such that* F *is differentiable on* $J_i \cap I$ *with continuous derivative at* x_i, $i = 0, \ldots, p-1$. *Then the inequality*

$$(2.2.10) \qquad \left|\frac{d}{dx} F^p(x_0)\right| > 1$$

implies that the orbit is a source.

Proof. By Theorem 2.2.4 and equality (2.2.8) each point x_i is a repeller for the dynamical system defined by F^p. Hence, the periodic orbit is a source.

Example 2.2.8 Let $F(x) = 3.6\,x(1 - x)$. The orbit $x_0 = \frac{2.3 + \sqrt{.69}}{3.6}$, $x_1 = \frac{2.3 - \sqrt{.69}}{3.6}$ is periodic of prime period 2 . From $F_x(x) = 3.6 - 7.2x$, we derive

$$F_x(x_0) = -1 - 2\sqrt{.69} \text{ and } F_x(x_1) = -1 + 2\sqrt{.69}.$$

Hence,

$$F^2{}_x(x_0) = (-1 - 2\sqrt{.69})(-1 + 2\sqrt{.69}) = 1 - 2.76 = -1.76.$$

The periodic orbit $\{x_0, x_1\}$ is a repeller.

Inequality (2.2.10) is certainly satisfied if $|F_x(x_i)| > 1$ for all $i = 0, 1, 2, \ldots, p-1$. However, it can happen that for some x_i we have $|F_x(x_i)| \le 1$, but (2.2.10) is still true. The following example illustrates this situation.

Example 2.2.9 Let $F : [0, 1] \to [0, 1]$ be defined by

$$F(x) = \begin{cases} 4x & 0 \le x < 1/4 \\ -3x + 7/4 & 1/4 \le x < 1/2 \\ -x/2 + 1/2 & 1/2 \le x \le 1 . \end{cases}$$

Then $F(1/6) = 2/3$ and $F(2/3) = 1/6$.

Hence, 1/6 is a periodic point of period 2. Moreover, $F_x(1/6) = 4$, while $|F_x(2/3)| = 1/2$. Since

$$|F_x(1/6)F_x(2/3)| = 2 > 1,$$

the periodic orbit is a repeller.

A careful analysis of the behavior of an orbit whose initial state is very close to 1/6 shows that in going from a neighborhood of 1/6 to a neighborhood of 2/3 the orbit gets further away from the periodic orbit, while in the successive state it gets closer. This alternating effect continues for a while. However, at the end of each two-step cycle, the state of the orbit is further away from the corresponding periodic state than it was before.

Analogous behavior can be found for an attracting periodic orbit $O(x_0, p)$ in the sense that we may have $|F_x(x_i)| > 1$ for some of the indices $i = 0, 1, ..., p$ even though

$$\left|\frac{d}{dx}F^p(x_0)\right| < 1.$$

Notice that the inequality holds for all points x_i of the orbit $O(x_0, p)$.

Example 2.2.10 Let $I = [0, 5/4]$ and $F : I \rightarrow I$ be defined by

$$F(x) = \begin{cases} 4x(1 - x) & 0 \leq x < 1/2 \\ 1 & 1/2 \leq x < 3/4 \\ 5/2 - 2x & 3/4 < x \leq 5/4\,. \end{cases}$$

Then $F(1/2) = 1$ and $F(1) = 1/2$. Since $F^2_x(1/2) = F^2_x(1) = 0$ the periodic orbit is an attractor. Notice that $|F_x(1)| = 2$.

Definition of weak repeller

Sometimes the derivative of the function governing a dynamical system satisfies a weaker inequality in an open and bounded interval J containing an equilibrium point x_s, in the sense that $|F_x(x)| > 1$ for all $x \in J \cap I$, except at $x = x_s$ where the derivative may not exist or $|F_x(x_s)| = 1$.

In this case we say that x_s is a **weak repeller**.

A similar definition holds for periodic orbits.

Example 2.2.11 Let $F : \mathbf{R} \to \mathbf{R}$ be defined by $F(x) = x + x^3$. The only equilibrium point of the dynamical system governed by F is $x_s = 0$. Since the derivative of F is $F_x(x) = 1 + 3x^2$, $F_x(0) = 1$ and $F_x(x) > 1$ for every $x \neq 0$, we see that the equilibrium point $x_s = 0$ is a weak repeller.

The orbits of points close to a weak repeller x_s separate from x_s and the distance from x_n to x_s increases as n gets larger. More precisely, let x_s be a weak repeller and $x_0 \in J \cap I$. As long as $x_i \in J \cap I$, $i = 0, 1, ..., p - 1$, we have

$$|x_n - x_s| > |x_{n-1} - x_s| > ... > |x_0 - x_s| \ .$$

The result follows from the Mean Value Theorem .

Instability

In the first chapter we introduced the notion of instability for a stationary state x_s or a periodic orbit of prime period p, $O(x_0, p)$. Repelling or weak repelling stationary states or periodic orbits are obviously unstable. The converse is not true as the following examples show.

Example 2.2.12 Let $F(x) = x - x^2$. The only equilibrium point is $x_s = 0$. For $0 < x < 1$ we have $|F_x(x)| < 1$ and $0 < F(x) < 1$. Every trajectory starting at a point $x_0 \in [0, 1]$ converges to 0.

For every $x < 0$ we have $F_x(x) > 1$ and $F(x) < 0$. The trajectory of every point $x_0 < 0$ goes to $-\infty$. The origin is an unstable point, but not a repeller or a weak repeller.

Example 2.2.13 Let

$$F(x) = \begin{cases} -2x - 1 & -\infty < x \leq -1 \\ -.5x + .5 & -1 < x \leq -.25 \\ -2.5x & -.25 < x \leq .25 \\ -.5x - .5 & .25 < x \leq 1 \\ -2x + 1 & 1 < x < \infty . \end{cases}$$

The interval $[-1, 1]$ is mapped onto itself by F, and $F(-1) = 1$, $F(1) = -1$. For every initial condition $x_0 \in [-1, 1]$, except $x_s = 0$, the trajectory $\{x_n\}$ converges to the periodic orbit $\{-1, 1\}$. The trajectory of every initial condition $|x_0| > 1$ goes to infinity.

Section 3. Global Behavior and Conjugacy

1. Global behavior

Until now we have confined our attention to local behavior of one-dimensional systems, in the sense that our analysis of stationary states and periodic orbits has been limited to the evolution of the system close to these points. We now want to gain some information on the **global behavior** of a dynamical system, at least in some simple cases. Similar situations on multi-dimensional systems are studied in Chapter 5.

Frequently a numerical investigation of a system seems to suggest that all its orbits converge to a stationary state or to a periodic orbit. This first impression however, may be misleading. A system may have more than one stationary state, and it may also have several, or even infinitely many, periodic orbits of different prime periods. In the dynamical evolution of the system, these stationary states and periodic orbits will neither converge to the equilibrium point nor to the periodic orbit where everything else seems to be going. To better understand this situation we first present a remarkable result which was established by the Russian mathematician A.N. Sarkovskii [Sarkovskii, 1964].

The formulation of his theorem requires that we first present the natural numbers $\mathbf{N} = \{1, 2, 3, ...\}$ in a different order. We shall call it Sarkovskii's ordering of $\mathbf{N}$.

The natural numbers are organized in a two dimensional array with infinitely many rows and columns. The first row contains all odd numbers, starting from 3, in increasing order. The second row contains, in the same order, the integers obtained by multiplying the elements of the first row by 2. The third row contains, in the same order, the integers obtained by multiplying the elements of the first row by $4 = 2^2$, etc. . The last row contains all powers of 2 in decreasing order, and ends with $2^2, 2, 1$.

3	5	7	9	11	13	15....................			
6	10	14	18	22	26	30....................			
12	20	28	36	44	52	60....................			
24	40	56	72	88..					
....									
....									
....			32	16	8	4	2	1.	

Every integer is included in this list, and appears in it exactly once .

Theorem 2.3.1 (Sarkovskii) *Let* I *be an interval and let* $F : I \to I$ *be continuous. Assume that* F *has a periodic orbit of period* p. *Then* F *has a periodic orbit of every period which follows* p *in Sarkovskii's ordering of* **N**.

From Theorem 2.3.1 we derive that a one-dimensional system with a periodic orbit of period 3 has a periodic orbit of every period. Moreover, the presence of a periodic orbit $O(x_0, p)$, $p > 1$, implies the presence of an equilibrium point. These results are certainly remarkable give the modest assumption from which they are derived.

When numerical evidence indicates that all orbits of a system converge to a periodic orbit $O(x_0, p)$, $p > 1$, Sarkovskii's theorem insures the presence of exceptions to this behavior.

Theorem 2.3.2 below gives sufficient conditions for the existence of a unique stationary state x_s and for the convergence to x_s of every orbit of the system. The result is fairly general and will be extended, with minor changes, to dynamical system in $\mathbf{R}^q$. The proof of the theorem is based on some fundamental results from Calculus, which we state without proof.

Recall that a sequence $\{x_n, n = 1, 2, ...\}$ is increasing (strictly increasing) if

$$x_n \leq x_{n+1} \qquad (x_n < x_{n+1}) \qquad n = 1, 2, \ldots .$$

Reversing the inequalities we obtain a decreasing (strictly decreasing) sequence. In some occasions the entire sequence does not satisfy the definition of increasing or decreasing, but if finitely many initial terms are deleted, then the sequence which is left does. In such cases we say that the original sequence is **eventually increasing** or **decreasing**.

Lemma 2.3.1 *Let* $\{x_n : n = 1, 2, 3, ...\}$ *be an eventually decreasing (increasing) sequence. Assume that* $\{x_n : n = 1, 2, 3, ...\}$ *is bounded below (above). Then the sequence is convergent, i. e. there exists a real number* w *such that*

$$x_n \to w \ \text{as} \ n \to +\infty .$$

Lemma 2.3.2 *Let* I *be an interval and* $F : I \to \mathbf{R}$ *be continuous on* I *and differentiable at every interior point of* I. *Assume that* $F_x(x) \geq 0$ *and it is zero only finitely* *many times on every bounded interval* J *contained in* I. *Then* F *is strictly increasing on* I.

We can now establish the following result.

Theorem 2.3.2 *Let* I *be an interval and* $F: I \to I$ *be continuous on* I *and differentiable at every interior point of* I. *Assume that*

- F *has a fixed point* x_s.
- $|F_x(x)| \leq 1$ *for all interior points* x *of* I.
- *In every bounded interval* $J \subset I$ *there are only finitely many points where* $|F_x| = 1$.

Then every orbit converges to x_s.

Proof. We establish first that F has only one fixed point and has no periodic orbits of prime period 2.

Define $G(x) = x - F(x)$ and notice that $G(x) = 0$ if and only if x is a fixed point of F. Hence $G(x_s) = 0$. Let us show that G satisfies the assumptions of Lemma 2.3.2.

The inequality $|F_x(x)| \leq 1$ implies that $G_x(x) = 1 - F_x(x)$ is non-negative. Moreover, $G_x(x) = 0$ if and only if $F_x(x) = 1$. Thus $G_x(x) = 0$ only finitely many times in every bounded interval. By Lemma 2.3.2, the function G is strictly increasing and x_s is the only point where G is zero. Consequently, x_s is the only fixed point of F.

To verify that F does not have periodic orbits of prime period 2 we follow a similar approach and we define $H(x) = x - F(F(x)) = x - F^2(x)$. Notice that $H(x) = 0$ if and only if $x = F(F(x))$. Hence $H(x_s) = 0$. From the Chain Rule we obtain

$$H_x(x) = 1 - F_x(y)F_x(x), \text{ with } y = F(x).$$

From $|F_x(x)| \leq 1$ it follows that $H_x(x) \geq 0$. Moreover $H_x(x) = 0$ only finitely many times in every bounded interval. By Lemma 2.3.2 H is strictly increasing and x_s is the only point where H is 0. Consequently, F does not have any periodic point of prime period 2.

We now establish that every trajectory converges to x_s.

First , notice that for every x_0 we have, by the Mean Value Inequality,

$$|x_1 - x_s| = |F(x_0) - x_s| = |F(x_0) - F(x_s)| \leq |x_0 - x_s| .$$

Hence, each state of a trajectory is at least as close to the fixed point as its predecessor.

Let $x_0 > x_s$. We have $G(x_0) > G(x_s) = 0$, since G is strictly increasing. This implies $x_1 = F(x_0) < x_0$. In general, we obtain

$$x_{n+1} < x_n \quad \text{if} \quad x_n > x_s$$

$$x_{n+1} > x_n \quad \text{if} \quad x_n < x_s .$$

From $|x_{n+1} - x_s| \le |x_n - x_s|$ we derive that successive states of the orbit to the right of x_s form a decreasing sequence bounded below by x_s, and successive states of the orbit to the left of x_s form an increasing sequence bounded above by x_s. By Lemma 2.3.1 the limit points of the orbit are either one, call it x_∞ , or two, namely x_L to the left of x_s, and x_R to the right of x_s. Thus, either $L(x_0) = \{x_\infty\}$ or $L(x_0) = \{x_L, x_R\}$.

Recall first that $L(x_0)$ has the important property $F(L(x_0)) = L(x_0)$. Hence, in the first case, we obtain that $F(x_\infty) = x_\infty$, i.e. x_∞ is a stationary state of F. Since F has only one stationary state, the equality $x_\infty = x_s$ follows. In the second case there are only two possibilities

- $F(x_L) = x_L$ and $F(x_R) = x_R$,

or

- $F(x_L) = x_R$ and $F(x_R) = x_L$.

The first alternative implies $x_L = x_R = x_s$. The second alternative is not acceptable, since F does not have any periodic point of prime period 2.
We have proved that every orbit converges to x_s.

In Theorem 2.3.2 we have assumed that F has a fixed point. A careful analysis of the proof shows that if I is a closed interval and there is a bounded trajectory then there is an equilibrium point. In other words, the trajectories of a map $F : I \to I$ such that

$$|F_x(x)| \le 1 \text{ and } |F_x(x)| = 1$$

only finitely many times in every bounded interval, are either all bounded or all unbounded. Therefore, once we have verified the condition on the derivative, we do not need to find if there is or there is not a fixed point. A sequence of iterates can tell us which case we are dealing with. If the sequence is bounded, then F has one and only one fixed point and every orbit converges to it. If the sequence is unbounded, then F does not have any fixed points and every orbit of F is unbounded.

Example 2.3.1 Let $F(x) = \text{ArcTan}\, x$. Then $x_s = 0$ is a stationary state. Moreover the derivative of ArcTanx is $1/(1 + x^2)$ and it satisfies the inequality $0 < 1/(1 + x^2) \leq 1$ with equality only at $x = 0$. Hence every orbit converges to 0 (see Fig. 2.3.1).

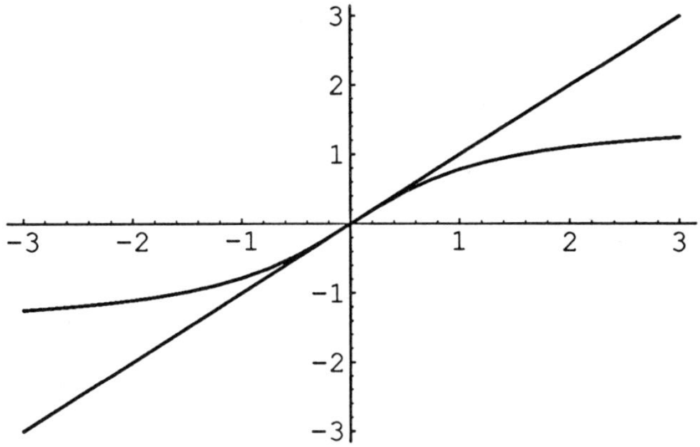

Fig. 2.3.1 The graph of ArcTan x and of the line $y = x$.

Example 2.3.2 Let $F(x) = \text{Cos}\, x$. There exists $x_s \in (0, 1)$ such that $F(x_s) = x_s$ (see Fig. 2.3.2). The function satisfies all assumptions of Theorem 2.3.2, since $|\text{Sin}\, x| \leq 1$ with equality only finitely many times in every bounded interval. Hence every orbit converges to x_s.

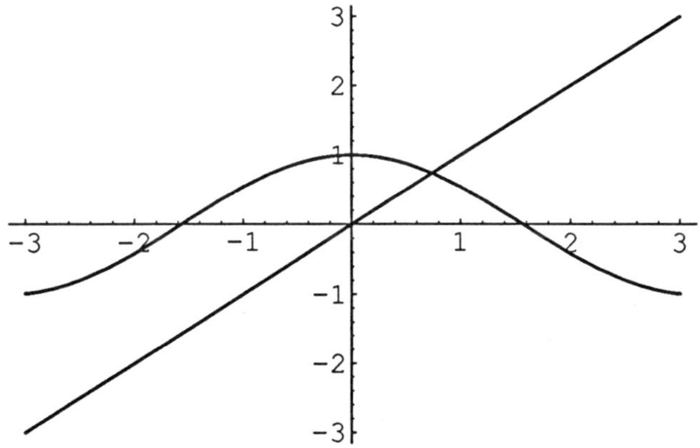

Fig. 2.3.2 The graph of the function Cos x and the line $y = x$.

Example 2.3.3 Let $F(x) = x + 2 - \text{ArcTan}\, x$. Then $F_x(x) = 1 - \frac{1}{1+x^2}$ and $0 \le F_x(x) < 1$, but F does not have any fixed point. Every trajectory of the dynamical system defined by F goes to infinity.

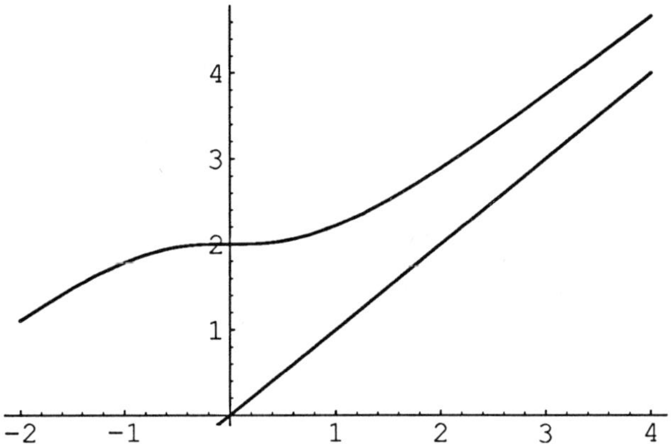

Fig. 2.3.3 Every orbit of the dynamical system generated by $F(x) = x + 2 - \text{ArcTan}\, x$ goes to infinity.

In many cases numerical evidence seems to indicate that all orbits of a dynamical system converge to a periodic orbit. The following example is typical.

Example 2.3.4 Let $F(x) = 3.2x(1 - x)$. We have seen that

$$x_0 = \frac{2.1 + \sqrt{.21}}{3.2}, \qquad x_1 = \frac{2.1 - \sqrt{.21}}{3.2}$$

is a periodic orbit of prime period 2 . Since

$$F^2{}_x(x_0) = (-1 - 2\sqrt{.21})(-1 + 2\sqrt{.21}) = 1 - .84 = .16$$

we know that the periodic orbit $\{x_0, x_1\}$ is an attractor. Numerical evidence suggests that every orbit converges to $\{x_0, x_1\}$. The exceptions, whose presence is insured by Sarkovskii's theorem, are O(0) and O(1).

There are no simple general criteria similar to the one provided by Theorem 2.3.2 to insure that a periodic orbit behaves like the orbit $\{x_0, x_1\}$ of the previous example.

2. More on conjugacy

Let $F : I \to I$ and $G : J \to J$ be conjugate via a map $h : I \to J$ and let x_s, $h(x_s)$ be two corresponding stationary states. The trajectory starting at $x_0 \in I$ converges to x_s if and only if the trajectory starting at $y_0 = h(x_0)$ converges to $h(x_s)$. This fact derives from the continuity of h and h^{-1}. In fact, since $F^n = h^{-1}G^n h$, we obtain

$$x_n = F^n(x_0) = h^{-1}G^n h(x_0) = h^{-1}G^n(y_0) = h^{-1}(y_n).$$

Hence the sequence $\{x_n\}$ converges to x_s if and only if the sequence $\{y_n\}$ converges to $h(x_s)$.

A similar argument can be used for periodic orbits of prime period p, since F^p and G^p are also conjugate via the same map h. It follows that a stationary state x_s or a periodic orbit $O(x_0, p)$ are stable or unstable if and only if the same is true for the corresponding stationary state $h(x_s)$ and periodic orbit $O(y_0, p)$. The limit set $L(x_0)$ of the orbit $O(x_0)$ is mapped by h in a one-to-one manner onto the limit set $L(y_0)$ of the orbit $O(y_0)$. Therefore $O(x_0)$ and $O(y_0)$ are either both finite or both infinite.

All these comments are based only on the continuity of h. Something more can be said when h is differentiable. For example, assume that $h_x(x_s) \neq 0$. From

$$G(h(x)) = h(F(x))$$

we derive

$$G_x(h(x_s))h_x(x_s) = h_x(F(x_s))F_x(x_s) = h_x(x_s)F_x(x_s).$$

Therefore, dividing by $h_x(x_s) \neq 0$, we obtain

(2.3.1) $$G_x(h(x_s)) = F_x(x_s) .$$

Similarly, from $G^p(h(x)) = h(F^p(x))$ we derive

$$(G^p)_x(h(x_0))h_x(x_0) = h_x(F^p(x_0))(F^p)_x(x_0) = h_x(x_0)(F^p)_x(x_0).$$

This implies

(2.3.2) $$(G^p)_x(h(x_0)) = (F^p)_x(x_0)$$

provided that $h_x(x_0) \neq 0$. Equality (2.3.1) guarantees that if x_s is a sink (source) the same is true for $h(x_s)$.

Equality (2.3.2) authorizes similar conclusions for the periodic orbit $O(x_0, p)$ and the corresponding orbit $O(h(x_0), p)$.

These comments, combined with the observations made at the beginning of this chapter show the importance of conjugacy in the study of dynamical systems.

Example 2.3.5 Recall that the two maps

$$F(x) = -2x^2 + 1 \text{ and } G(x) = -2|x| + 1$$

are conjugate in $[-1, 1]$ via the map $h(x) = \frac{2}{\pi}\text{ArcSin}x$.

Noticc that $|G_x(x)| = 2$ except at $x = 0$ where G is not differentiable. The trajectory starting from $x_0 = 0$ is eventually stationary, since $x_1 = 1$, $x_2 = x_3 = ... = -1$. The map G has the periodic orbit of period three

$$x_0 = -\frac{5}{9} \qquad x_1 = \frac{1}{9} \qquad x_2 = \frac{7}{9} .$$

Therefore, according to the theorem of Sarkovskii, the dynamical system defined by G has a periodic orbit of every period. All these orbits are repellers. Hence the same is true for all stationary states and periodic orbits of F. A direct analysis of F is not as simple since $|F_x(x)| < 1$ for $|x| < .25$.

Example 2.3.6 One can verify that the maps $T(x) = 4x^3 - 3x$ and the map

$$L(x) = \begin{cases} 3x + 2 & \text{if } -1 \le x \le -\frac{1}{3} \\ -3x & \text{if } -\frac{1}{3} < x \le \frac{1}{3} \\ 3x - 2 & \text{if } \frac{1}{3} < x \le 1 \end{cases}$$

with $x \in [-1, 1]$ are conjugate via the same map h of the previous example.

Since $|L_x(x)| = 3$ at every point where L is differentiable, we obtain that all stationary states and periodic orbits of the dynamical system induced by L in $[-1, 1]$ are repellers. Consequently, all stationary states and periodic orbits of the dynamical system induced by T in the same interval are repellers.

Section 4. Parameter Space Analysis

1. Bifurcation points and bifurcation diagrams

Let a be a scalar and $x_{n+1} = F(x_n, a)$ be a **one-parameter family** of dynamical systems governed by a function F which is differentiable with continuous derivatives with respect to x and a. We assume that for every fixed value of the parameter a, the action of the map $F(., a)$ takes every point of a certain region R_a (the domain of $F(., a)$ or some subset of it) into itself. As the parameter a ranges through the values of an interval I, the region R_a may change.

Let $J \subset I$ be an interval with more than one point and $\alpha : J \to \mathbf{R}$ be a continuous function such that for every $a \in J$, $\alpha(a)$ has the property

$$F(\alpha(a), a) = \alpha(a),$$

i.e. $\alpha(a)$ is a stationary state of $F(., a)$ for every $a \in J$. Then the map α (and its graph) is called a **branch** of fixed points of (the one-parameter family) F. Notice that as a ranges in J, the dynamical system $F(., a)$ changes.
In a similar manner we define branches of periodic orbits of prime period p.

Example 2.4.1 Let $F(x, a) = ax(1 - x)$, $x \in [0, 1]$, $a \in [0, 4]$. From $F(x, a) = x$ we derive $x = 0$ and $x = \frac{a - 1}{a}$. Therefore, the function $\alpha : [0, 4] \to \mathbf{R}$ defined by $\alpha(a) = 0$ is a branch of fixed points of F. Notice that the fixed point $\alpha(a)$ is independent of the parameter a. The function $\alpha : [1, 4] \to \mathbf{R}$ defined by $\alpha(a) = \frac{a - 1}{a}$ is another branch of fixed points of F. This time the fixed point changes with a.

Example 2.4.2 Recall that the function of the previous example does not have any periodic orbit of prime period 2 for $a < 3$ because the discriminant $a^2 - 2a - 3$ is negative. For $a > 3$ there are exactly two periodic points of prime period 2, namely

$$x_0 = \frac{a + 1 + \sqrt{a^2 - 2a - 3}}{2a}, \qquad x_1 = \frac{a + 1 - \sqrt{a^2 - 2a - 3}}{2a}.$$

Thus, we have two branches of periodic points of prime period two of F when the parameter $a \in [3, 4]$. They are given by the functions

$$\alpha_1(a) = \frac{a + 1 + \sqrt{a^2 - 2a - 3}}{2a} \quad \text{and} \quad \alpha_2(a) = \frac{a + 1 - \sqrt{a^2 - 2a - 3}}{2a}.$$

We have included the value $a = 3$ in the interval J. Notice that at $a = 3$ the two periodic points coincide.

A value $a_0 \in I$ will be called a **bifurcation point** of the one-parameter family F (or, briefly, of F) if the following two conditions are verified :

- there are two different functions $\alpha_i : J_i \to \mathbf{R}$, $i = 1, 2$ whose graphs are branches of stationary states or periodic points of F and $\alpha_1(a_0) = \alpha_2(a_0)$;

- $J_1 \cap J_2 \neq a_0$ and $\alpha_1(a) \neq \alpha_2(a)$, $a \in J_1 \cap J_2$, $a \neq a_0$.

We identify $\mathbf{R}^2$ with a plane and we represent the parameter a along the horizontal axis. The pictorial display obtained by plotting all branches of stationary states and periodic points is called **bifurcation diagram** of F on I.

Example 2.4.3 Let $F(x, a) = ax - x^3$. We can set $R_a = \mathbf{R}$ for every value of a. We can also imagine that the parameter a assumes every value in $\mathbf{R}$.

The stationary states of F(., a) are the solutions of the equation $x = ax - x^3$. We see that $x_0(a) = 0$ is always a stationary state of our system, independently of a. No other solution exists for $a \leq 1$. When the parameter a crosses the value 1, two more stationary states appear, namely

$$x_1(a) = \sqrt{a - 1} \text{ and } x_2(a) = - \sqrt{a - 1}.$$

At $a = 2$, two periodic orbits of period 2 appear (see a more detailed analysis later in this section). Moreover, the periodic orbit of period two

$$y_1(a) = \sqrt{a+1} \quad \text{and} \quad y_2 = - \sqrt{a+1}$$

exists for every $a > -1$.

The bifurcation diagram looks as in Fig. 2.4.1 .

The points $a = -1$, $a = 1$ and $a = 2$ are bifurcation points of the one-parameter family of maps $\{F(., a)\colon a \in \mathbf{R}\}$.

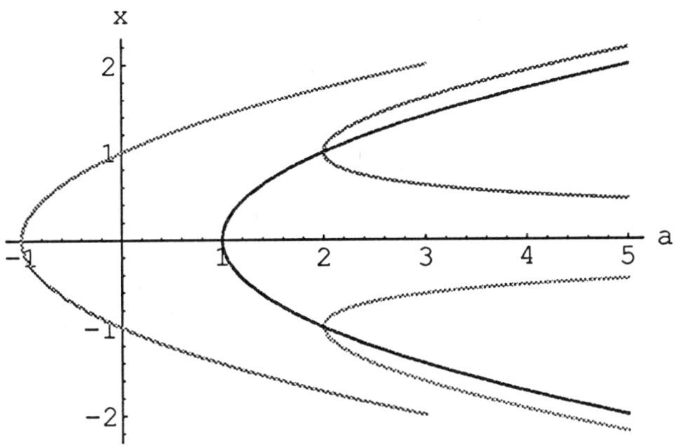

Fig. 2.4.1 The bifurcation diagram for the map $F(x, a) = ax - x^3$.

2. **Period-doubling bifurcation**

We now illustrate with examples four different types of bifurcation points, starting with the so-called **period-doubling** . We also analyze their stability.

Example 2.4.4 Let $F(x, a) = ax(1 - x)$, $0 \le a \le 4$. The interval $[0, 1]$ is mapped into itself by $F(., a)$ for every value of $a \in [0, 4]$. The stationary state $x_s(a) = \frac{a-1}{a}$ is a sink for every value of a in the interval $(1, 3)$. At $a = 3$ we have

$$x_s(3) = 2/3 \quad \text{and} \quad F_x(2/3, 3) = -1.$$

For $a > 3$ the point $x_s(a) = \frac{a-1}{a}$ becomes a repeller and a periodic orbit of period 2 appears,

$$x_1(a) = \frac{a + 1 + \sqrt{a^2 - 2a - 3}}{2a}, \quad x_2(a) = \frac{a + 1 - \sqrt{a^2 - 2a - 3}}{2a}.$$

The orbit is a sink for all values of $a \in (3, 1 + \sqrt{6})$. In fact

$$F_x(x_1(a), a)\, F_x(x_2(a), a) = 4 + 2a - a^2.$$

Setting $a = 3 + h, h > 0,$ we obtain

$$F_x(x_1(3 + h), 3 + h)\, F_x(x_2(3 + h), 3 + h) = 1 - 4h - h^2 = f(h).$$

The function f is decreasing for $h > 0$ and the periodic orbit becomes a repeller after the value of h for which

$$1 - 4h - h^2 = -1$$

or $h = -2 + \sqrt{6}$. This implies $a = 1 + \sqrt{6}$. For $a > 1 + \sqrt{6}$ the system F(., a) has a periodic orbit of prime period 4. This orbit is a sink for $a \in (1 + \sqrt{6}, a_3)$, $a_3 \approx 3.54409..$.

This behavior continues and the sequence $a_1 = 3$, $a_2 = 1+ \sqrt{6}$, a_3, ... converges to $a_\infty \approx 3.569946$, at which value of the parameter a the system has aperiodic trajectories.

The elements of the sequence $\{a_n\}$ are called **period-doubling bifurcation points** of the one-parameter family of maps $\{F(x, a) = ax(1 - x) : a \in [0, 4]\}$. The entire sequence $\{a_n\}$ is sometimes called a **cascade of bifurcations.**

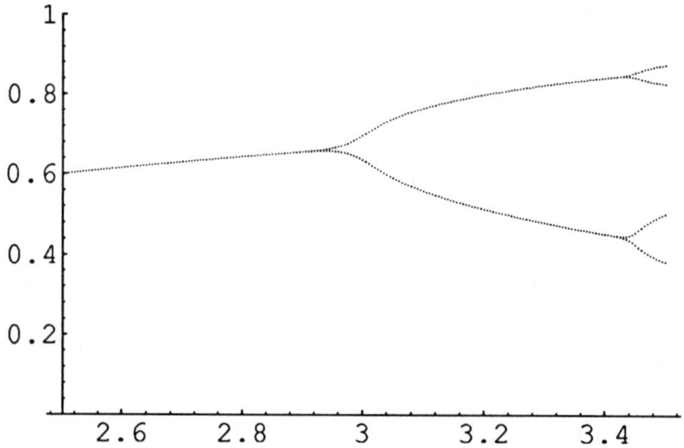

Fig. 2.4.2 Part of the bifurcation diagram of the one-parameter family of maps $F(x, a) = ax(1 - x)$. The equilibrium point $(a - 1)/a$ loses its stability for $a > 3$, and an attractive periodic orbit of prime period 2 arises. For $a > 1 + \sqrt{6}$ the orbit becomes a repeller and its attractive character is passed on to a periodic orbit of prime period 4 etc..

The sequence $\{a_k\}$ converges to a_∞ according to the relation

(2.4.1) $$a_k \approx a_\infty - c\alpha^{-k}$$

where $c = 2.6327...$, $\alpha = 4.669201609...$.

In 1975 M.J. Feigenbaum [Feigenbaum, 1975] noted that this period-doubling cascade of bifurcations appears in many other processes and that the constant α is the same. Therefore α has been called *Feigenbaum*'s *constant.* The knowledge of a_k for k large enough allows us to estimate a_∞ using (2.4.1). For example, with the aid of a simple computer program, we can easily find that for $F(x, a) = ax(1 - x)$ we have

(2.4.2)
$$a_1 = 3 \qquad a_5 = 3.568759..$$
$$a_2 = 1 + \sqrt{6} \qquad a_6 = 3.569692..$$
$$a_3 = 3.544090.. \qquad a_7 = 3.569891..$$
$$a_4 = 3.564407.. \qquad a_8 = 3.569934...$$

From (2.4.1) and (2.4.2) we can estimate a_∞ using the approximate equality

$$a_\infty \approx \frac{\alpha a_n - a_{n-1}}{\alpha - 1}.$$

Moreover, we derive that the constant α is the limit of the sequence

(2.4.3) $$\frac{a_n - a_{n-1}}{a_{n+1} - a_n}.$$

The sequence of period-doubling bifurcations starts with two branches $b_{21}(a)$, $b_{22}(a)$ of periodic points of prime period 2 arising from a branch $b_1(a)$ of stationary states when the parameter a crosses a value a_1. The derivative $F_x(x, a)$ computed at the point $(b_1(a_1), a_1)$ is equal to -1, i.e.

$$F_x(b_1(a_1), a_1) = -1.$$

The stationary states of the branch $b_1(a)$ are attractors when the parameter a ranges in an open interval $(a_1 - \varepsilon_1, a_1)$. For $a > a_1$ the attracting feature is passed on to the periodic orbit of period 2

$$x_1(a) = b_{21}(a), \qquad x_2(a) = b_{22}(a).$$

This orbit remains attracting until the parameter reaches a certain value a_2.

At $a = a_2$ two branches of periodic points of prime period $4 = 2^2$ start from $b_{21}(a)$ and two more from $b_{22}(a)$. Denote the four branches by

$$b_{41}(a), b_{42}(a) \text{ from } b_{21}(a), \text{ and } b_{43}(a), b_{44}(a) \text{ from } b_{22}(a).$$

For $a > a_2$ the four points

$$x_{4i}(a) = b_{4i}(a), \qquad i = 1, 2, 3, 4,$$

form a periodic orbit of period 4. The derivative of $F^2(x, a)$ computed at

$$a = a_2 \text{ and } x_1(a) = b_{21}(a) \text{ or } x_2(a) = b_{22}(a),$$

is equal to -1. The periodic orbit of period 2 becomes a repeller for $a > a_2$. The periodic orbit of period 4 is a sink until the parameter reaches a value a_3.

At this point the bifurcation scheme is repeated with $F^4(x, a)$ replacing $F^2(x, a)$ and $F^8(x, a)$ replacing $F^4(x, a)$ until the parameter reaches a value a_4. The process just described continues as a goes through the sequence of period-doubling bifurcation points $\{a_n\}$.

3. Pitchfork bifurcation

An analysis of the diagram presented in Fig. 2.4.2 reveals the presence of another type of bifurcation point. The figure, produced using the map $F(x, a) = ax(1 - x)$, can also be read in terms of the second iterate F^2 of F.

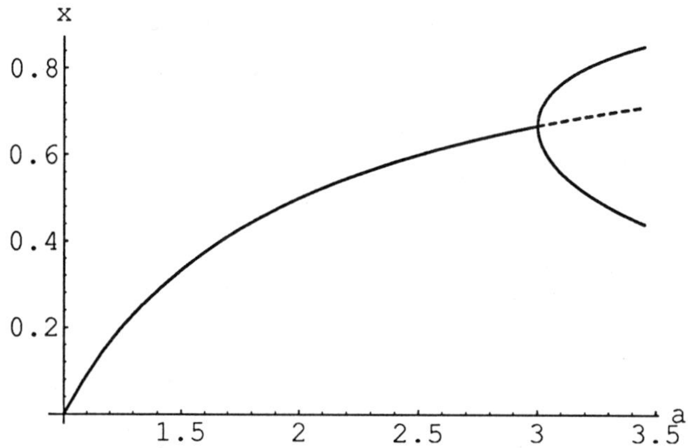

Fig. 2.4.3 The stationary state $(a - 1)/a$ of the system governed by the second iterate of $F(x, a) = ax(1 - x)$ loses stability at $a = 3$, and a pair of stable equilibrium points of F^2 arises. The value $a_p = 3$ is called a pitchfork bifurcation for $F^2(x, a)$.

The stationary state $(a - 1)/a$ of F is a fixed point for F^2. The periodic orbit of period 2 which arises at $a = 3$, becomes now a pair of fixed points of F^2. Hence, when a increases through 3, the fixed point $(a - 1)/a$ loses stability, and two branches of stable fixed points of F^2 arise. The derivative of F^2 computed at $a = 3$ and $x_1(3) = 2/3$ is equal to 1.

The value $a_p = 3$ is called **pitchfork bifurcation** point for the one-parameter family of maps $\{F^2(x, a) = ax(1 - x)(1 - ax(1 - x)) : a \in [0, 4]\}$.

Notice the difference between the period-doubling and the pitchfork. In both cases the stationary state loses its stability, which in the period-doubling is passed on to a periodic trajectory of period 2, while in the pitchfork is passed on to a pair of fixed points. The derivative of the map is

-1 at a period-doubling bifurcation point,
1 at a pitchfork bifurcation point.

Another example of period-doubling and pitchfork bifurcation is provided by the map of Example 2.4.3.

Example 2.4.5 Let $F(x, a) = ax - x^3$. We have seen that $x_s = 0$ is always a stationary state of our system, independently of a. No other stationary state exists for $a \leq 1$. When the parameter crosses the value 1, two more stationary states appear, namely $x_1(a) = \sqrt{a-1}$ and $x_2(a) = -\sqrt{a-1}$ (see Fig. 2.4.4).

From $F_x(x, a) = a - 3x^2$ we derive that 0 is a repeller for $|a| > 1$ and a sink for $|a| < 1$. At $a = 1$ we have $F_x(0, 1) = 1$. The value $a_p = 1$ is a pitchfork bifurcation point for the one-parameter family of maps.

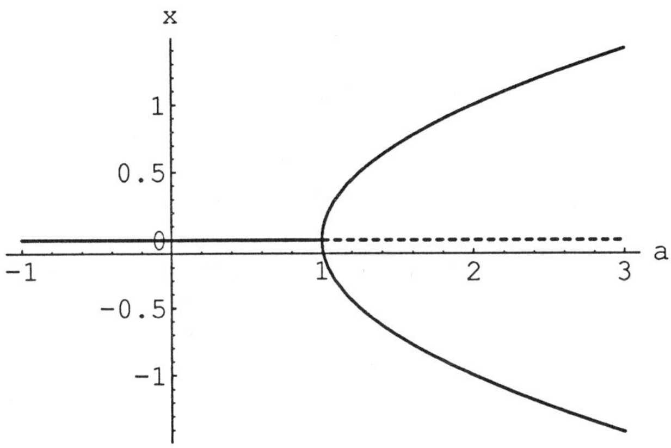

Fig. 2.4.4 The bifurcation diagram for $F(x, a) = ax - x^3$. As a crosses the value 1 the stationary state $x = 0$ becomes a repeller. Its stability is passed on to the pair of stationary states $\sqrt{a-1}$ and $-\sqrt{a-1}$.

As a crosses the value 1, the stability of the branch $x_s(a) = 0$ is passed on to the stationary states $x_1(a)$ and $x_2(a)$. In fact

$$F_x(\sqrt{a-1}, a) = F_x(-\sqrt{a-1}, a) = 3 - 2a, \text{ and } -1 < 3 - 2a < 1 \text{ as long as } 1 < a < 2 .$$

The two stationary states are repellers for $a > 2$, and their stability is passed on to a pair of periodic orbits of period 2 . At $a = 2$ we have

$$F_x(1, 2) = F_x(-1, 2) = -1 .$$

The parameter value $a_d = 2$ is a period-doubling bifurcation point.

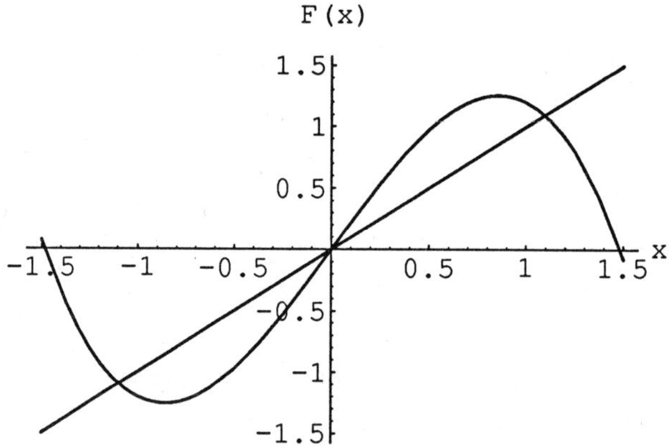

Fig. 2.4.5 The graph of $F(x, 2.2) = 2.2x - x^3$ suggests that $a = 2$ is a period-doubling bifurcation point.

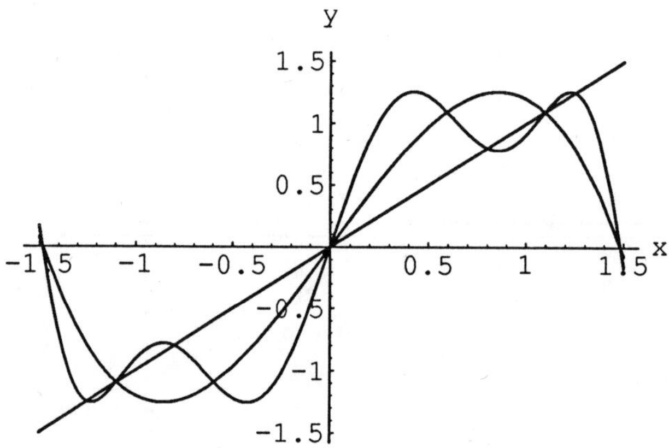

Fig. 2.4.6 The graph of the second iterate of $F(x, 2.2) = 2.2x - x^3$ in the interval [-1.5, 1.5] confirms that the value $a = 2$ is a period-doubling bifurcation point.

In fact, for $a > 2$ there are two pairs of periodic orbits of period 2, given by the following formula

$$x_i(a) = \pm\sqrt{\frac{a \pm \sqrt{a^2 - 4}}{2}}, \qquad i = 1, 2, 3, 4.$$

Both periodic orbits are attractors as long as $2 < a < \sqrt{5}$.

The graph of the second iterate $F(F(x, 2.2), 2.2)$ in the interval [-1.5, 1.5] (see Fig. 2.4.6) shows that one orbit is close to the stationary state $-\sqrt{a - 1}$ and the other is close to the stationary state $\sqrt{a - 1}$.

4. Transcritical bifurcation

The one-parameter family of maps $F(x, a) = ax(1 - x)$ provides an example of another type of bifurcation, called **transcritical.** As mentioned before, the branch of stationary states $x_0(a) = 0$ is present for every value of the parameter a. A second branch of stationary states is given by $x_1(a) = \frac{a - 1}{a}$. The two branches meet at $a = 1$.

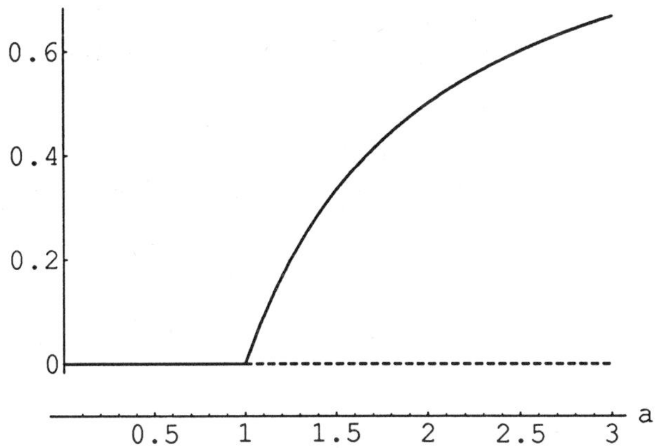

Fig. 2.4.7 The bifurcation of a stable branch of fixed points at $a = 1$ for the logistic map $ax(1 - x)$. The stationary state $x = 0$ loses its stability.

For $a \in [0, 1)$ the stationary state $x_1(a)$ is a repeller since

$$F_x\left(\frac{a - 1}{a}, a\right) = 2 - a,$$

while $x_0(a) = 0$ is a sink. For $a > 1$ the roles are reversed. The equilibrium $x_0(a) = 0$ is a source for all values of $a \in (1, 4]$, while $x_1(a) = \frac{a - 1}{a}$ is a sink for $a \in (1, 3)$.

The value $a_t = 1$ is called a **transcritical bifurcation point**. The part of the branch $\frac{a-1}{a}$ which comes before $a = 1$ is usually neglected in the bifurcation analysis since the corresponding stationary state is negative, and therefore irrelevant to the dynamical process modeled by the one-parameter family of maps (frequently called logistic).

In general, transcritical bifurcation takes place at a value a_t of the parameter when two branches of stationary states exist in an open interval J containing a_t and they meet at $a = a_t$. An exchange of stability among the two branches takes place at $a = a_t$.

5. Fold Bifurcation

A parameter value a_f is said to be a **fold bifurcation point**, if the following two conditions are satisfied.

- There are two branches α_1 and α_2 of stationary states or periodic orbits of the same prime period which are defined on a common interval J which begins (or ends) at a_f.

- $\alpha_1(a_f) = \alpha_2(a_f)$, $\alpha_1(a) \neq \alpha_2(a)$, for $a \in J$, $a \neq a_f$, and no other branch α has the property $\alpha(a_f) = \alpha_1(a_f)$.

Example 2.4.6 Let $F(x, a) = (x + a^2 - 1)(x^2 - 2x - a) + x$. The stationary states are the solutions of the equation

$$(x + a^2 - 1)(x^2 - 2x - a) = 0,$$

namely

$$1 - a^2,\ 1 + \sqrt{1 + a}\ ,\text{ and }\ 1 - \sqrt{1 + a}\,.$$

The solutions are plotted in Fig. 2.4.8 as functions of the parameter a.

We see that for $a < -1$ there is only one equilibrium point, namely $x_s(a) = 1 - a^2$, which is unstable. At the point $a = -1$ two branches of stationary states appear,

$$x_1(a) = 1+ \sqrt{1+a} \quad\text{and}\quad x_2(a) = 1- \sqrt{1+a}\ ,$$

and their starting point is different from $x_s(-1) = 0$.

The upper branch is unstable and the lower is stable. The parameter value $a_f = -1$ is a fold bifurcation point. Notice that $F_x(1, -1) = 1$.

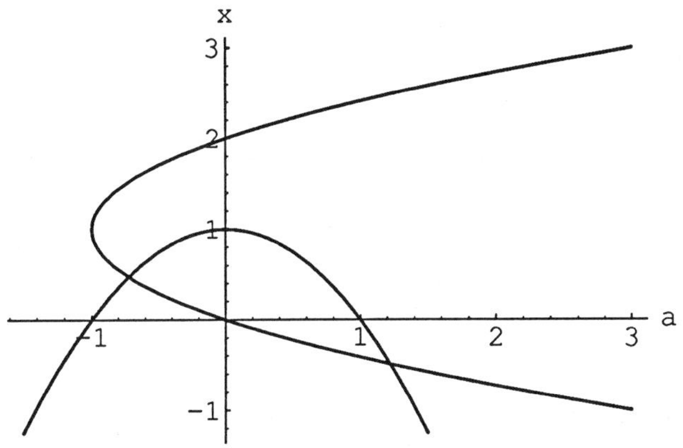

Fig. 2.4.8 At $a_f = -1$ the dynamical system governed by $F(x, a) = (x + a^2 - 1)(x^2 - 2x - a) + x$ has a fold bifurcation.

Necessary and sufficient conditions can be given for the existence of a period-doubling, pitchfork, transcritical and fold bifurcation at a stationary state x_b and for a specific value a_b of the parameter a. In the first case, we must have $F_x(x_b, a_b) = -1$. In the other three cases we must have $F_x(x_b, a_b) = 1$. The sufficient conditions, called transversality conditions, involve successive derivatives of F.

Instead of listing them here we prefer to encourage the reader to use a careful numerical investigation to detect the type of bifurcation process present in the dynamical system under study. The complexity of many processes frequently makes the intelligent use of numerical techniques a good source of prompt and reliable information. We conclude this section with some additional examples.

Example 2.4.7 Let $F(x, a) = x^2 - a$. The stationary states of F are the solutions of the equation $x^2 - a = x$ or

$$x_1(a) = \frac{1 + \sqrt{1 + 4a}}{2}, \quad x_2(a) = \frac{1 - \sqrt{1 + 4a}}{2}.$$

These two branches of stationary states exist for $a > -\frac{1}{4}$. At $a_f = -\frac{1}{4}$ we have $x_1 = x_2$. Hence a_f is a fold bifurcation (see Fig. 2.4.9). The derivative of F at $(x_1(a_f), a_f)$ is

$$F_x(0.5, -0.25) = 1.$$

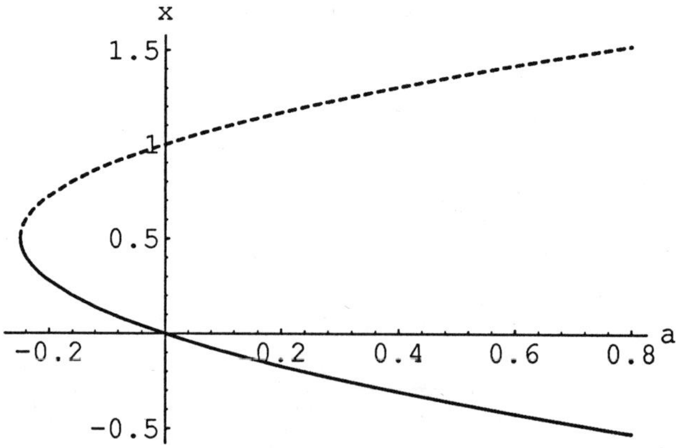

Fig. 2.4.9. The supercritical fold bifurcation for the one-parameter family $F(x, a) = x^2 - a$ at $a_f = -.25$. The upper branch is unstable and the lower is stable

Example 2.4.8 Let $F(x, a) = x^2 + a$. The stationary states of F are the solutions of the equation $x^2 + a = x$ or

$$x_1 = \frac{1 + \sqrt{1 - 4a}}{2}, \quad x_2 = \frac{1 - \sqrt{1 - 4a}}{2} .$$

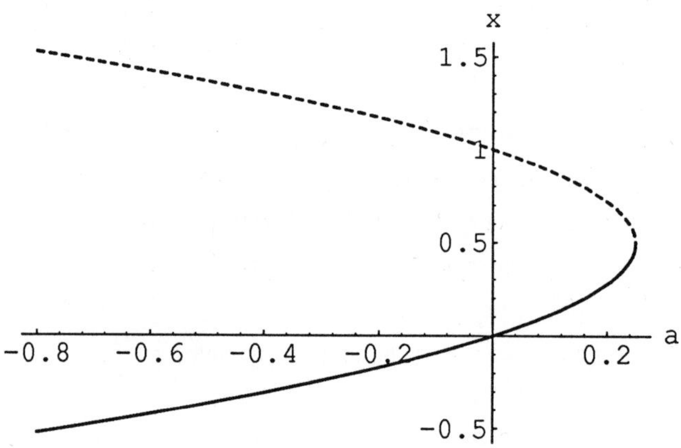

Fig. 2.4.10. The subcritical fold bifurcation for the map $F(x, a) = x^2 + a$ at $a_f = .25$.

The two branches of stationary states exist for $a < \frac{1}{4}$. The upper branch is unstable and the lower is stable. At $a_f = \frac{1}{4}$ we have $x_1 = x_2$. Hence, a_f is a fold bifurcation. However, since the branches disappear as the parameter crosses the bifurcation value, we call .25 a **subcritical fold** (see Fig. 2.4.10). The value -.25 of Example 2.4.7 is called a **supercritical fold.**

Example 2.4.9 Consider the one-parameter family of maps $G(x, a) = x^3 - ax$. Notice that $x_s(a) = 0$ is always a fixed point regardless of the value of a. For $a > -1$ there are two more stationary states, namely $x_1(a) = \sqrt{a+1}$ and $x_2(a) = -\sqrt{a+1}$ which are never stable. In fact their stability requires $a \in (-2, -1)$, but they exist only for $a \geq -1$. The point $a_p = -1$ is a pitchfork bifurcation point (see Fig. 2.4.11).

The stationary state $x_s(a) = 0$ is a repeller for $|a| > 1$ and a sink for $a \in (-1, 1)$.

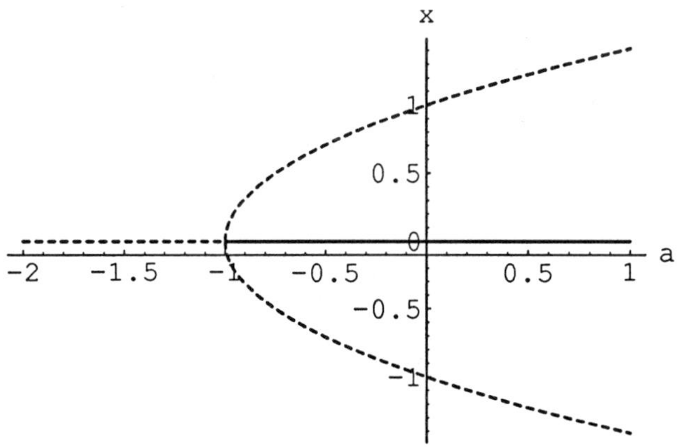

Fig. 2.4.11 The pitchfork bifurcation at $a = -1$ for the map $G(x, a) = x^3 - ax$. The stationary state $x_s = 0$ becomes a sink as a crosses the value -1.

Example 2.4.10 Let $H(x, a) = ax + x^3$. The one-parameter family of maps has the branches of stationary states

$$x_1(a) = \sqrt{1 - a} \quad \text{and} \quad x_2(a) = -\sqrt{1 - a}$$

for $a < 1$. The equilibrium points of the two branches are always repellers. The $x_0(a) = 0$ equilibrium is a repeller for $|a| > 1$ and a sink for $a \in (-1, 1)$. The value $a_p = 1$ is a pitchfork bifurcation. The bifurcation diagram is " symmetric" to the one of G with respect to the vertical axis (see Fig. 2.4.12).

In the case of Example 2.4.9 we call $a_p = -1$ a **supercritical pitchfork.** In the present example the value $a_p = 1$ is called a **subcritical pitchfork.**

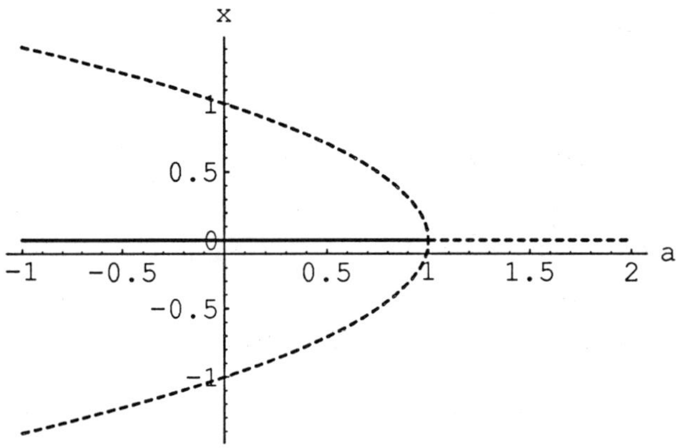

Fig. 2.4.12 The subcritical pitchfork bifurcation at $a_p = 1$ for the map $H(x, a) = x^3 + ax$. The stationary state $x_s = 0$ is a repeller for $a > 1$.

Example 2.4.11 The one-parameter family of maps $K(x, a) = -x^3 - ax$ gives a bifurcation diagram which is symmetric with respect to the vertical axis to the one of $F(x, a) = ax - x^3$ of Example 2.4.5 (see Fig. 2.4.4 and Fig. 2.4.13). The value $a_p = -1$ is a subcritical pitchfork.

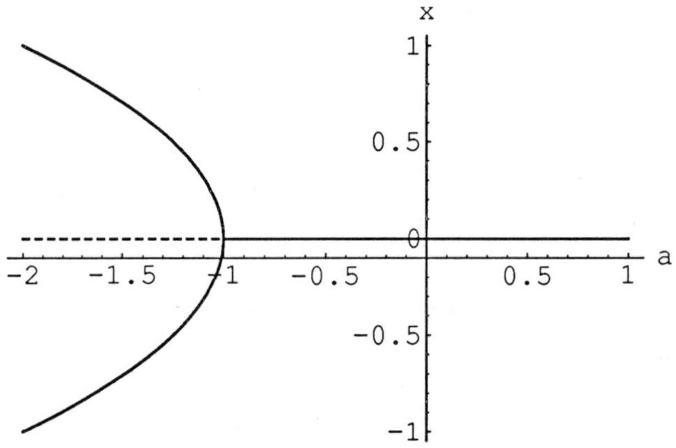

Fig. 2.4.13 The subcritical pitchfork bifurcation at $a_p = -1$ for the one-parameter family $K(x, a) = -x^3 - ax$. The stationary state $x_s = 0$ is a sink for $a \in (-1, 1)$.

Problems

Section 1.

2.1.1 Show that the two systems

$$x_{n+1} = x_n(a - bx_n)\,,\ x \in [0, \tfrac{a}{b}] \ \text{and} \ y_{n+1} = ay_n(1 - y_n),\ y \in [0,1]$$

are topologically conjugate (Hint : set $y = \frac{b}{a}x$).

2.1.2 From the results established for the dynamical system (2.1.2) derive the corresponding results for the dynamical system (2.1.3) using the conjugacy suggested in the previous problem.

2.1.3 Verify that the two systems

$$x_{n+1} = 4x_n(1 - x_n)\,,\ x \in [0, 1] \ \text{and} \ y_{n+1} = -2y_n^2 + 1,\ y \in [-1, 1]$$

are conjugate .

2.1.4 Verify that the two systems $y_{n+1} = -2\,|y_n| + 1$, $y \in [-1, 1]$ and $x_{n+1} = F(x_n)$,

$x \in [0,1]$, with $F(x) = \begin{cases} 2x & \text{if } 0 \le x \le \frac{1}{2} \\ 2 - 2x & \text{if } \frac{1}{2} < x \le 1 \end{cases}$ are conjugate.

2.1.5 Show that the dynamical system generated by $F(x, a) = x(a - bx)$ changes the interval $I_a = [0, a/b]$ into itself, provided that $0 \le a \le 4$.

2.1.6 Show that the map $F(x) = 2x^3 - x$ maps the interval $[-1, 1]$ into itself.

2.1.7 Use the cobweb method to verify that given $x_0 = 4$ the sequence $\{x_n = \text{Cos}\ x_{n-1} : n = 1, 2, \ldots\}$ converges to the unique stationary state $x_s = \text{Cos}\ x_s$.

2.1.8 Use the cobweb method to verify that given $x_0 = 4$, the sequence $\{x_n = \text{Log}(1+ x_{n-1}),\ n = 1, 2, \ldots\}$ converges to 0 (use natural log).

2.1.9 Find h^{-1}, where $h(x) = \frac{2}{\pi}\,\text{ArcSin}x$.

2.1.10 Let $F(x) = -2x^2 + 1$, $x_0 = -\text{Sin}(\pi/10)$, $x_1 = \text{Sin}(3\pi/10)$. Verify that

$$F(x_0) = x_1 \text{ and } F(x_1) = x_0.$$

Section 2.

2.2.1 Discuss the stationary states of the system $F(x, a, b) = ax^3 - bx$, as functions of the parameters a and b.

2.2.2 Assume that $x \in I$ has the property

$$x = F^2(x) \text{ and } x = F^3(x).$$

Prove that $x = F(x)$. In general prove that from

$$x = F^n(x) \text{ and } x = F^m(x),\ n > m,$$

we derive $x = F^{n-m}(x)$.

2.2.3 Let $F : [0, 1] \to [0, 1]$ be defined by

$$F(x) = \begin{cases} 2x & 0 \le x \le 0.5 \\ 2(1 - x) & 0.5 < x \le 1\,. \end{cases}$$

Find a periodic orbit of prime period 2 of F.

2.2.4 Let $F(x, a) = -ax^2 + 1$. Find the values of the parameter a for which the function maps the interval $[-1, 1]$ into itself. Among these values find the ones for which F has a periodic orbit of prime period 2. Determine for which values of a the periodic orbit is a sink.

2.2.5 Let $F(x, a) = -a|x| + 1$. Find the values of a for which the function maps $[-1, 1]$ into itself. Find those values of a for which the map has a periodic orbit of prime period 2. Determine the stability of the orbit.

2.2.6 Let $F(x) = -2|x| + 1$. Plot the graph of F and the graph of F^2. Derive from the two graphs the number and location of the periodic orbits of prime period 2 of F.

Section 3.

2.3.1 Prove that every natural number appears once and exactly once in the Sarkovskii ordering of **N** (Hint: recall that every integer can be uniquely factorized as a product of prime numbers).

2.3.2 Using the Intermediate Value Theorem prove that if $F : I \to I$ has a periodic point of period 2, then F has a fixed point.

2.3.3 Construct a dynamical system with a periodic orbit of period 5, but not a periodic orbit of period 3.

2.3.4 Prove that the equation $x = \mathrm{Cos}\, x$ has one and only one solution.

2.3.5 Prove that every trajectory of the dynamical system governed by the function $F(x) = \mathrm{Sin}\, x$ converges to 0.

2.3.6 Let $F : [0, 1] \to [0, 1]$ be such that $F_x(x) > 0$ and $F_{xx}(x) > 0$.
Prove that $F_x(0) < 1$. (Hint : if $F_x(0) \geq 1$ then $F_x(x) > 1$ for all $x \in (0, 1)$. Use the Mean Value Theorem to show that F(1) is larger than 1).

2.3.7 Let F be as in Problem 2.3.6. Assume that $F(0) = 0$. Prove that every orbit with initial state $x_0 \in (0,1)$ converges to 0.

Section 4.

2.4.1 Study the bifurcation diagram of the map $F(x, a) = a - x^2$. In particular determine the fold bifurcation and if there is any period-doubling bifurcation point. Investigate the stability of the equilibrium points and periodic orbits of period 2.

2.4.2 Study the bifurcation diagram of the map $F(x, a) = - a + x^2$. Determine the fold bifurcation and if there is any period-doubling bifurcation point. Investigate the stability of the equilibrium points and periodic orbits of period 2.

2.4.3 Study the bifurcation diagram of the map $F(x,a) = ax - x^2$. In particular determine the transcritical bifurcation, and if there is any period-doubling bifurcation point. Investigate the stability of the equilibrium points and periodic orbits of period 2.

2.4.4 Study the bifurcation diagram of the map $F(x, a) = - ax + x^2$. In particular determine the transcritical bifurcation, and if there is any period-doubling bifurcation point. Investigate the stability of the equilibrium points and periodic orbits of period 2.

2.4.5 Study the bifurcation diagram of the map $F(x, a) = - ax + x^3$. In particular determine the pitchfork bifurcation, and if there is any period-doubling bifurcation point.

2.4.6 Study the bifurcation diagram of the map $F(x, a) = (x + a^2 - 1)(x^2 - 2x - a) + x$. In particular investigate the stability of the branches of fixed points.

2.4.7 Let $F(x, a) = ax^3 - (a - 1)x$. Study the bifurcation diagram of the map F. In particular verify that $x = -1$, $x = 0$, and $x = 1$ are always stationary states. Study their stability. Find the period-doubling bifurcation points and analyze the stability of the bifurcating branches.

CHAPTER 3

$\mathbf{R}^q$, FUNCTIONS AND MATRICES

SUMMARY

The purpose of this chapter is to present a brief survey of the fundamental results of Calculus of Several Variables and Linear Algebra, which will be needed in the study of the evolution of linear and nonlinear discrete dynamical systems in a dimension higher than one, and of their chaotic behavior. In section 1 we introduce the q-dimensional space $\mathbf{R}^q$ *and its main properties, in section 2 we present the basic facts about square matrices, in section 3 we study the action in* $\mathbf{R}^q$ *of matrices of size* q, *and in section 4 we examine nonlinear differentiable maps in* $\mathbf{R}^q$.

Section 1 . Structure of $\mathbf{R}^q$

1. Definition of $\mathbf{R}^q$

The **elements** of $\mathbf{R}^q$ are ordered q-tuples of real numbers . Each q-tuple is written as follows : $(x_1, x_2, ..., x_q)$, with $x_i \in \mathbf{R}$, $i = 1, ..., q$. The qualification ordered means that q-tuples with the same entries, in different order, are considered different q-tuples. For example in $\mathbf{R}^3$ the triples $(1, 2, 3)$ and $(3, 2, 1)$ are different.

The elements of $\mathbf{R}^q$ are often called **vectors** or **points** and the q entries of each vector are called **components** or **coordinates** . Bold-face characters, such as $\mathbf{x}$, $\mathbf{a}$, will denote vectors. The vector $\mathbf{0} = (0, 0, ..., 0)$ is called **zero vector** or **origin**. Other frequently used vectors are

$$\mathbf{e}_1 = (1, 0, 0, ..., 0), \mathbf{e}_2 = (0, 1, 0, ..., 0), ..., \mathbf{e}_q = (0, 0, ..., 0, 1) .$$

We shall denote by E the set of these vectors. Hence $E = \{\mathbf{e}_1, \mathbf{e}_2, ..., \mathbf{e}_q\}$.

Three operations are defined on the elements of $\mathbf{R}^q$.

Addition. To every pair of vectors of $\mathbf{R}^q$,

$$\mathbf{x} = (x_1, x_2, ..., x_q) \text{ and } \mathbf{y} = (y_1, y_2, ..., y_q),$$

we associate the vector

$$\mathbf{z} = \mathbf{x} + \mathbf{y} = (x_1 + y_1,\ x_2 + y_2, ..., x_q + y_q).$$

Multiplication by a scalar. To each vector $\mathbf{x}$ and real number r we associate the vector

$$\mathbf{y} = r\mathbf{x} = (rx_1, rx_2, ..., rx_q).$$

Inner Product. To each pair of vectors $\mathbf{x}$ and $\mathbf{y}$ of $\mathbf{R}^q$ we associate the real number $\mathbf{x}\bullet\mathbf{y}$ defined as follows

$$\mathbf{x}\bullet\mathbf{y} = x_1y_1 + x_2y_2 + ... + x_qy_q.$$

The inner product is sometimes called **scalar product** or **dot product.**

Example 3.1.1 Let $\mathbf{x} = (1, 2, 3)$, $\mathbf{y} = (3, 1, 4)$ and $r = -2$. Then

$$\mathbf{z} = \mathbf{x} + \mathbf{y} = (1 + 3, 2 + 1, 3 + 4) = (4, 3, 7) ;$$

$$r\mathbf{x} = (-2, -4, -6) ;$$
$$\mathbf{x}\bullet\mathbf{y} = 1\times 3 + 2\times 1 + 3\times 4 = 17 .$$

The three operations listed above have several properties, some particular to each single operation, and others involving two or even all three operations. We provide below a list of some of these properties. The ones listed, and the others not included here, are derived from corresponding properties of the real numbers.

Let $\mathbf{x}, \mathbf{y}, \mathbf{z}$ be vectors of $\mathbf{R}^q$ and a,b be real numbers. Then

1. $\mathbf{x}\bullet\mathbf{y} = \mathbf{y}\bullet\mathbf{x}$;
2. $\mathbf{x}\bullet\mathbf{x} \geq 0$ and $\mathbf{x}\bullet\mathbf{x} = 0$ if and only if $\mathbf{x} = \mathbf{0}$;
3. $\mathbf{x}\bullet(a\mathbf{y} + b\mathbf{z}) = a\mathbf{x}\bullet\mathbf{y} + b\mathbf{x}\bullet\mathbf{z}$;
4. $0 \leq (a\mathbf{y} - b\mathbf{z})\bullet(a\mathbf{y} - b\mathbf{z}) = a^2\mathbf{y}\bullet\mathbf{y} - 2ab\mathbf{y}\bullet\mathbf{z} + b^2\mathbf{z}\bullet\mathbf{z}$.

Property 1 is based on the commutative property of the product between real numbers, since

$$\mathbf{x}\bullet\mathbf{y} = x_1y_1 + x_2y_2 + \dots + x_ny_n = y_1x_1 + y_2x_2 + \dots + y_qx_q = \mathbf{y}\bullet\mathbf{x} .$$

Property 2 is evident, since

$$\mathbf{x}\bullet\mathbf{x} = x_1^2 + x_2^2 + \dots + x_q^2 .$$

Property 3 follows from the distributive property of the product with respect to the sum, the commutative property of the product and the sum, and the associative property of the sum. In fact

$$\mathbf{x}\bullet(a\mathbf{y} + b\mathbf{z}) = x_1(ay_1 + bz_1) + \dots + x_q(ay_q + bz_q) =$$
$$ax_1y_1 + bx_1z_1 + \dots + ax_qy_q + bx_qz_q =$$
$$ax_1y_1 + \dots + ax_qy_q + bx_1z_1 + \dots + bx_qz_q = a\mathbf{x}\bullet\mathbf{y} + b\mathbf{x}\bullet\mathbf{z}.$$

Property 4 follows from 1, 2 and 3. In fact, from Property 2 we have

$$0 \leq (a\mathbf{y} - b\mathbf{z})\bullet(a\mathbf{y} - b\mathbf{z}),$$

which is the left hand side inequality. From Property 3 we have

$$(a\mathbf{y} - b\mathbf{z})\bullet(a\mathbf{y} - b\mathbf{z}) = (a\mathbf{y} - b\mathbf{z})\bullet(a\mathbf{y}) + (a\mathbf{y} - b\mathbf{z})\bullet(- b\mathbf{z}).$$

By Property 1 and 3 we obtain

$$(ay - bz)\bullet(ay) = ay\bullet(ay - bz) = a^2\, \mathbf{y}\bullet\mathbf{y} - ab\mathbf{y}\bullet\mathbf{z}\,,$$

and

$$(ay - bz)\bullet(-bz) = (-bz)\bullet(ay - bz) = -\,ab\mathbf{y}\bullet\mathbf{z} + b^2\mathbf{z}\bullet\mathbf{z}\,.$$

Thus

$$(ay - bz)\bullet(ay - bz) = a^2\mathbf{y}\bullet\mathbf{y} - 2ab\mathbf{y}\bullet\mathbf{z} + b^2\mathbf{z}\bullet\mathbf{z}\,.$$

From Property 4 with a = 1, we have

$$0 \le \mathbf{y}\bullet\mathbf{y} - 2b\mathbf{y}\bullet\mathbf{z} + b^2\mathbf{z}\bullet\mathbf{z}$$

for every pair of vectors **y** and **z** and for every real number b. Hence, assuming $\mathbf{z} \neq \mathbf{0}$, the minimum of the function

$$f(b) = \mathbf{y}\bullet\mathbf{y} - 2b\mathbf{y}\bullet\mathbf{z} + b^2\mathbf{z}\bullet\mathbf{z}$$

cannot be negative. Setting the derivative $f_b(b) = -2\mathbf{y}\bullet\mathbf{z} + 2b\mathbf{z}\bullet\mathbf{z}$ equal to 0 we obtain the value $b = \frac{\mathbf{y}\bullet\mathbf{z}}{\mathbf{z}\bullet\mathbf{z}}$. The second derivative of f is always positive. Hence $b = \frac{\mathbf{y}\bullet\mathbf{z}}{\mathbf{z}\bullet\mathbf{z}}$ is the absolute minimum b_m of f. Since $f(b_m)$ cannot be negative, we get

$$f(\,b_m) = \mathbf{y}\bullet\mathbf{y} - \frac{(\mathbf{y}\bullet\mathbf{z})^2}{\mathbf{z}\bullet\mathbf{z}} \ge 0\,.$$

We have just established the basic inequality, called the **Cauchy-Schwarz Inequality,**

(3.1.1) $$(\mathbf{y}\bullet\mathbf{z})^2 \le (\,\mathbf{y}\bullet\mathbf{y})(\mathbf{z}\bullet\mathbf{z})\,.$$

Property 2 is used to define the Euclidean **norm** or **length** of a vector **x**, denoted by $\|\mathbf{x}\|$,

(3.1.2) $$\|\mathbf{x}\| = \sqrt{\mathbf{x}\bullet\mathbf{x}}\,.$$

The Euclidean **norm** is a function from $\mathbf{R}^q$ into $\mathbf{R}$, $\|\,.\,\| : \mathbf{R}^q \to \mathbf{R}$, and it satisfies the following properties

i) $\|\mathbf{x}\| \ge 0$ and $\|\mathbf{x}\| = 0$ if and only if $\mathbf{x} = \mathbf{0}$;

ii) $\|r\mathbf{x}\| = |r|\,\|\mathbf{x}\|$;

iii) $|\,\|\mathbf{x}\| - \|\mathbf{y}\|\,| \le \|\mathbf{x} + \mathbf{y}\| \le \|\mathbf{x}\| + \|\mathbf{y}\|$.

Inequalities i) and ii) are obvious. Inequality iii) needs justification. First, we rewrite the Cauchy-Schwarz Inequality using the norm

(3.1.3) $$(\mathbf{x}\bullet\mathbf{y})^2 \le \|\mathbf{x}\|^2\,\|\mathbf{y}\|^2 \quad \text{or} \quad |\mathbf{x}\bullet\mathbf{y}| \le \|\mathbf{x}\|\,\|\mathbf{y}\|.$$

Then, we observe that from (3.1.3) and Property 4, one obtains

$$\|\mathbf{x}+\mathbf{y}\|^2 = \|\mathbf{x}\|^2 + 2\,\mathbf{x}\bullet\mathbf{y} + \|\mathbf{y}\|^2 \le \|\mathbf{x}\|^2 + 2\,\|\mathbf{x}\|\,\|\mathbf{y}\| + \|\mathbf{y}\|^2 = (\,\|\mathbf{x}\| + \|\mathbf{y}\|)^2 \,.$$

This establishes the right hand side part of iii). Similarly, from (3.1.3) once more, we derive

$$(\|\mathbf{x}\| - \|\mathbf{y}\|)^2 = \|\mathbf{x}\|^2 - 2\,\|\mathbf{x}\|\,\|\mathbf{y}\| + \|\mathbf{y}\|^2 \le \|\mathbf{x}\|^2 + 2\,\mathbf{x}\bullet\mathbf{y} + \|\mathbf{y}\|^2 = \|\mathbf{x}+\mathbf{y}\|^2.$$

This establishes the left hand side part of iii).

From (3.1.3) we obtain $-1 \le \dfrac{\mathbf{x}\bullet\mathbf{y}}{\|\mathbf{x}\|\,\|\mathbf{y}\|} \le 1$. It is therefore legitimate to define the angle θ between the two vectors $\mathbf{x}$ and $\mathbf{y}$ as the one such that

(3.1.4) $$\text{Cos}\,\theta = \frac{\mathbf{x}\bullet\mathbf{y}}{\|\mathbf{x}\|\,\|\mathbf{y}\|} \,,$$

and to say that the two vectors are **orthogonal** if $\text{Cos}\,\theta = 0$.
The vectors of $E = \{\mathbf{e}_1, ..., \mathbf{e}_q\}$ are pair-wise orthogonal.

Example 3.1.2 Let $\mathbf{x} = (1, 2, 3)$, $\mathbf{y} = (3, 1, 4)$. Then $\|\mathbf{x}\| = \sqrt{14}$, $\|\mathbf{y}\| = \sqrt{26}$. We find

$$17 = |\mathbf{x}\bullet\mathbf{y}| \le \sqrt{14}\,\sqrt{26} = \|\mathbf{x}\|\,\|\mathbf{y}\| \approx 19\,;$$

$$|\,\|\mathbf{x}\| - \|\mathbf{y}\|\,| = \sqrt{26} - \sqrt{14} \approx 1.35 \le \|\mathbf{x}+\mathbf{y}\| = \sqrt{74} \approx 8.6 \le \|\mathbf{x}\| + \|\mathbf{y}\| \approx 8.8 \;;$$

$$\text{Cos}\,\theta = \frac{17}{\sqrt{14}\sqrt{26}} \approx .89, \qquad \theta \approx 27^{o}.$$

Every function from $\mathbf{R}^q$ into $\mathbf{R}$ which satisfies i), ii), iii) is called a norm in $\mathbf{R}^q$. For example, the taxi-cab norm we mentioned in Chapter 1. Another is the maximum norm defined by $\|\mathbf{x}\|_{max} = \max\{|x_i| : i = 1, 2, ..., q\}$. Every pair of norms, $\|.\|_1$, $\|.\|_2$ in $\mathbf{R}^q$ are equivalent, in the sense that there are two constants k, K (depending on the two norms), such that

$$k\|\mathbf{x}\|_1 \le \|\mathbf{x}\|_2 \le K\|\mathbf{x}\|_1$$

for every $\mathbf{x} \in \mathbf{R}^q$. For this reason all results of this book, although stated usually in terms of the Euclidean norm, are valid, with minor adjustments, in any other norm.

The Euclidean norm is used to define the **Euclidean distance** between $\mathbf{x}$ and $\mathbf{y}$

$$(3.1.5) \qquad d(\mathbf{x}, \mathbf{y}) = \|\mathbf{x} - \mathbf{y}\| .$$

The following properties of the distance are derived from the properties of the norm

j) $d(\mathbf{x}, \mathbf{y}) \geq 0$ and $d(\mathbf{x}, \mathbf{y}) = 0$ if and only if $\mathbf{x} = \mathbf{y}$;

jj) $d(\mathbf{x}, \mathbf{y}) = d(\mathbf{y}, \mathbf{x})$;

jjj) $d(\mathbf{x}, \mathbf{y}) \leq d(\mathbf{x}, \mathbf{z}) + d(\mathbf{z}, \mathbf{y})$.

The **ball** B, **sphere** S, and **the closed ball** D, centered at $\mathbf{x}$ and with radius r, are defined as follows:

- $B(\mathbf{x}, r) = \{\mathbf{y} \in \mathbf{R}^q : \|\mathbf{x} - \mathbf{y}\| < r\}$;
- $S(\mathbf{x}, r) = \{\mathbf{y} \in \mathbf{R}^q : \|\mathbf{x} - \mathbf{y}\| = r\}$;
- $D(\mathbf{x}, r) = \{\mathbf{y} \in \mathbf{R}^q : \|\mathbf{x} - \mathbf{y}\| \leq r\}$.

When $\mathbf{x} = \mathbf{0}$ we frequently omit it. Thus, for example, $D(r) = D(\mathbf{0}, r)$.

Example 3.1.3 Let $\mathbf{x} = (2, 1)$, $r = 3$. Then

$$B(\mathbf{x}, 3) = \{\mathbf{y} \in \mathbf{R}^2,\ \mathbf{y} = (x, y) : \sqrt{(2 - x)^2 + (1 - y)^2} < 3\ \} ;$$

$$S(\mathbf{x}, 3) = \{\mathbf{y} \in \mathbf{R}^2,\ \mathbf{y} = (x, y) : \sqrt{(2 - x)^2 + (1 - y)^2} = 3\ \} ;$$

$$D(\mathbf{x}, 3) = \{\mathbf{y} \in \mathbf{R}^2,\ \mathbf{y} = (x, y) : \sqrt{(2 - x)^2 + (1 - y)^2} \leq 3\ \} .$$

A subset U of $\mathbf{R}^q$ is called **open** if for every element $\mathbf{x} \in U$ there is a positive real number r such that $B(\mathbf{x}, r) \subset U$. A subset C of $\mathbf{R}^q$ is said to be **closed** if its complement is open.

The **boundary** ∂A of a set $A \subset \mathbf{R}^q$ is the set of points $\mathbf{x} \in \mathbf{R}^q$ such that for every $r > 0$ the ball $B(\mathbf{x}, r)$ intersects both A and its complement.

The **closure** of a set A is the union of A and its boundary. It is denoted by $\bar{A}$. The **interior** $\overset{o}{A}$ of a set A is $\overset{o}{A} = \{\mathbf{x} \in A : \mathbf{x} \notin \partial A\}$. Thus,

$$S(\mathbf{x}, r) = \partial D(\mathbf{x}, r) = \partial B(\mathbf{x}, r),\quad \bar{B}(\mathbf{x}, r) = D(\mathbf{x}, r),\quad \overset{o}{D}(\mathbf{x}, r) = B(\mathbf{x}, r).$$

2. Linear combination and linear independence

Let $\mathbf{x}_1, ..., \mathbf{x}_p$ be p vectors in $\mathbf{R}^q$ and let $a_1, ..., a_p$ be p real numbers. The vector

$$\mathbf{x} = a_1\mathbf{x}_1 + ... + a_p\mathbf{x}_p$$

is called a **linear combination** of the vectors $\mathbf{x}_1, ..., \mathbf{x}_p$. The set of all possible linear combinations of $\mathbf{x}_1, ..., \mathbf{x}_p$ is called the **span** of $\mathbf{x}_1, ..., \mathbf{x}_p$ and will be denoted by Span$\{\mathbf{x}_1, ..., \mathbf{x}_p\}$. $\mathbf{R}^q$ = Span$\{\mathbf{e}_1, ..., \mathbf{e}_q\}$ since every vector $\mathbf{x} = (x_1, ..., x_q)$ can be written as a linear combination of the elements of E : $\mathbf{x} = x_1\mathbf{e}_1 + ... + x_q\mathbf{e}_q$.

Notice that $\mathbf{0} \in$ Span$\{\mathbf{x}_1, ..., \mathbf{x}_p\}$, with $a_1 = ... = a_p = 0$. Whenever this is the only possible choice for obtaining the vector $\mathbf{0}$ as a linear combination of $\mathbf{x}_1, ..., \mathbf{x}_p$, we say that the vectors $\mathbf{x}_1, ..., \mathbf{x}_p$ are **linearly independent.** When the vector $\mathbf{0}$ can be obtained as a linear combination of $\mathbf{x}_1, ..., \mathbf{x}_p$ in some other way (i.e. with some of the coefficients $a_1, ..., a_p$ different from 0), we say that $\mathbf{x}_1, ..., \mathbf{x}_p$ are **linearly dependent.**

It is easily seen that the vectors of $E = \{\mathbf{e}_1, ..., \mathbf{e}_q\}$ are linearly independent. More generally, p vectors pair-wise orthogonal and different from $\mathbf{0}$, are linearly independent. In fact, let $\mathbf{x}_1, ..., \mathbf{x}_p$ be p pair-wise orthogonal vectors, all different from the zero vector. Assume

(3.1.6) $$a_1\mathbf{x}_1 + ... + a_p\mathbf{x}_p = \mathbf{0}\ .$$

Taking the inner product with $\mathbf{x}_1$ we obtain

(3.1.7) $$\mathbf{x}_1\bullet(a_1\mathbf{x}_1 + ... + a_p\mathbf{x}_p) = \mathbf{x}_1\bullet\ \mathbf{0} = 0\ .$$

Since the vectors are pair-wise orthogonal we have $\mathbf{x}_1\bullet\mathbf{x}_i = 0$ for $i = 2, 3, ..., p$. These equalities and (3.1.7) imply

$$\mathbf{x}_1\bullet(a_1\mathbf{x}_1 + ... + a_p\mathbf{x}_p) = a_1\|\mathbf{x}_1\|^2 = 0.$$

Hence $a_1 = 0$. In a similar manner we prove that all coefficients are 0. Therefore, the vectors are linearly independent.

Example 3.1.4 Let us check if the vectors (1, 1, 1), (-1, 1, 0) and (1, 2, -1) are linearly independent.

We set

$$a_1(1, 1, 1) + a_2(-1, 1, 0) + a_3 (1, 2, -1) = (0, 0, 0).$$

This implies

$$a_1 - a_2 + a_3 = 0, \quad a_1 + a_2 + 2 a_3 = 0, \quad a_1 - a_3 = 0.$$

The vectors will be linearly dependent if this system of three equations in three unknowns has a non zero solution. From the third equation we derive $a_1 = a_3$. Substitution into the other two gives $2a_1 = a_2$ and $3a_1 = - a_2$. Therefore, $5a_1 = 0$. The only solution is $a_1 = a_2 = a_3 = 0$. The three vectors are linearly independent.

Example 3.1.5 Let $\mathbf{x}_1 = (1, 0, 1)$, $\mathbf{x}_2 = (2, 1, 3)$, $\mathbf{x}_3 = (-1, -1, -2)$. From

$$a_1\mathbf{x}_1 + a_2\mathbf{x}_2 + a_3\mathbf{x}_3 = \mathbf{0},$$

we now derive

$$a_1 + 2a_2 - a_3 = 0, \qquad a_2 - a_3 = 0, \qquad a_1 + 3a_2 - 2a_3 = 0.$$

From the second equation we obtain $a_2 = a_3$. Substituting this equality into the first or the third leads to $a_1 + a_2 = 0$. Hence, every solution of the system can be written in the form $(-a_2, a_2, a_2)$ for every possible value of a_2. For example, for $a_2 = 1$ we obtain

$$-\mathbf{x}_1 + \mathbf{x}_2 + \mathbf{x}_3 = \mathbf{0} .$$

The three vectors are linearly dependent.

Recall that p vectors in $\mathbf{R}^q$ are always linearly dependent whenever $p > q$, since the homogeneous system in the unknowns $a_1, ..., a_p$ has more variables than equations.

3. Subspaces, basis and dimension

A subset **X** of $\mathbf{R}^q$ is said to be a **subspace** if for every pair of vectors **x, y** in **X**, and for every pair of real numbers a, b the vector $\mathbf{z} = a\mathbf{x} + b\mathbf{y}$ is again in **X**. Since we can always choose $a = b = 0$ we see that the vector **0** must belong to every subspace **X**. $\mathbf{R}^q$ is obviously a subspace of itself. The vector **0** is a subspace of $\mathbf{R}^q$.

We have previously defined $\text{Span}\{\mathbf{x}_1, ..., \mathbf{x}_p\}$. This is a subspace of $\mathbf{R}^q$. In fact if $\mathbf{x}, \mathbf{y} \in \text{Span}\{\mathbf{x}_1, ..., \mathbf{x}_p\}$, then **x** and **y** are linear combinations of $\mathbf{x}_1, ..., \mathbf{x}_p$ and we can write

$$\mathbf{x} = a_1\mathbf{x}_1 + ... + a_p\mathbf{x}_p ,$$
$$\mathbf{y} = b_1\mathbf{x}_1 + ... + b_p\mathbf{x}_p .$$

Thus,

$$a\mathbf{x} + b\mathbf{y} = a(a_1\mathbf{x}_1 + ... + a_p\mathbf{x}_p) + b(b_1\mathbf{x}_1 + ... + b_p\mathbf{x}_p) = (aa_1 + bb_1)\mathbf{x}_1 + ... + (aa_p + bb_p)\mathbf{x}_p ,$$

which is again a linear combination of $\mathbf{x}_1, ..., \mathbf{x}_p$. Therefore it belongs to $\text{Span}\{\mathbf{x}_1, ..., \mathbf{x}_p\}$.

Example 3.1.6 Let $\mathbf{x}_1 = (1, 0, 1)$, and $\mathbf{x}_2 = (0, 1, 1)$. $\mathbf{X} = \text{Span}\{\mathbf{x}_1, \mathbf{x}_2\}$ is the subspace of $\mathbf{R}^3$ of those vectors $\mathbf{x}$ having the form $\mathbf{x} = a\mathbf{x}_1 + b\mathbf{x}_2 = (a, b, a+b)$. We see that the third component is the sum of the other two. Denoting the components by x, y, z we obtain that $\mathbf{X}$ is characterized by the equation $z = x + y$.

A **basis** of a subspace $\mathbf{X}$ is any set of linearly independent vectors $\mathbf{x}_1, ..., \mathbf{x}_m$ such that $\mathbf{X} = \text{Span}\{\mathbf{x}_1, ..., \mathbf{x}_m\}$. Since $\mathbf{R}^q = \text{Span}\{\mathbf{e}_1, ..., \mathbf{e}_q\}$ and the vectors of E are linearly independent, we see that $\mathbf{e}_1, \mathbf{e}_2, ..., \mathbf{e}_q$ is a basis of $\mathbf{R}^q$. It is usually called the **standard** or **canonical basis** of $\mathbf{R}^q$.

It can be shown that if a basis of $\mathbf{X}$ has m vectors, then every other basis of $\mathbf{X}$ has exactly m vectors. The number m is called the **dimension** of $\mathbf{X}$ and it is denoted by dimX. Accordingly, $\mathbf{R}^q$ has dimension q.

Example 3.1.7 Let $\mathbf{x} = (1, 1, 2)$. $\text{Span}\{(1, 1, 2)\}$ is the subspace of all vectors of the form $\{(a, a, 2a) : a \in \mathbf{R}\}$. Its dimension is 1.

Let $\mathbf{X}$ be the set of all vectors of $\mathbf{R}^3$ which are orthogonal to $\mathbf{x} = (1, 1, 2)$. A vector $\mathbf{y} = (y_1, y_2, y_3) \in \mathbf{X}$ if and only if $\mathbf{x}\bullet\mathbf{y} = 0$, i.e. $y_1 + y_2 + 2y_3 = 0$. The two linearly independent vectors $\mathbf{u} = (-1, 1, 0)$ and $\mathbf{v} = (2, 0, -1)$ belong to $\mathbf{X}$.

From $y_1 = - y_2 - 2y_3$ we obtain

$$\mathbf{y} = (- y_2 - 2y_3, y_2, y_3) = y_2(-1, 1, 0) - y_3(2, 0, -1) = y_2\mathbf{u} - y_3\mathbf{v}.$$

Hence, every vector of $\mathbf{X}$ is a linear combination of the two vectors $\mathbf{u}$ and $\mathbf{v}$, which implies $\mathbf{X} = \text{Span}\{\mathbf{u}, \mathbf{v}\}$. Since $\mathbf{u}$ and $\mathbf{v}$ are linearly independent, they are a basis of $\mathbf{X}$ and $\dim \mathbf{X} = 2$.

A basis made of pair-wise orthogonal vectors is called an **orthogonal basis.**

When the pair-wise orthogonal vectors have length one, the basis is called **orthonormal**. Thus, E is an orthonormal basis of $\mathbf{R}^q$.

Example 3.1.8 Let $\mathbf{X}$ be the two-dimensional subspace of the previous example. An orthogonal basis for $\mathbf{X}$ can be selected as follows. Choose any vector $\mathbf{u}_1 \in \mathbf{X}, \mathbf{u}_1 \neq \mathbf{0}$. For example, let $\mathbf{u}_1 = (-1, 1, 0)$. A second vector $\mathbf{u}_2$ must belong to $\mathbf{X}$ and it must be orthogonal to $\mathbf{u}_1$. Hence, denoting by (a, b, c) the components of $\mathbf{u}_2$ we must have

$$-a + b = 0 \qquad \text{(for } \mathbf{u}_2 \text{ to be orthogonal to } \mathbf{u}_1\text{);}$$

$$a + b + 2c = 0 \qquad \text{(for } \mathbf{u}_2 \text{ to be in } \mathbf{X}\text{).}$$

From the first equation we get $b = a$, and from the second $c = -a$. With $a = 1$, we obtain $\mathbf{u}_2 = (1, 1, -1)$. The two vectors form an orthogonal basis of $\mathbf{X}$. To get an orthonormal basis we divide each vector by its norm. Since $\|\mathbf{u}_1\| = \sqrt{2}$ and $\|\mathbf{u}_2\| = \sqrt{3}$ we obtain the orthonormal basis

$$\mathbf{v}_1 = \mathbf{u}_1/\sqrt{2}\ , \quad \mathbf{v}_2 = \mathbf{u}_2/\sqrt{3}\,.$$

4. **Functions and continuity**

A function $F : \mathbf{R}^q \to \mathbf{R}^p$ can be viewed as an action on the elements of $\mathbf{R}^q$, which moves each element of $\mathbf{R}^q$ to an element of $\mathbf{R}^p$.

Example 3.1.9 Let $F(x, y, z) = (x - y,\ y^2, y + z^2, x + z)$. F is a function from $\mathbf{R}^3$ into $\mathbf{R}^4$ and $F(1, 1, 1) = (0, 1, 2, 2)$, $F(2, -1, 0) = (3, 1, -1, 2)$ etc..

As we see from the previous example, F is given by its **component functions** from $\mathbf{R}^q$ into $\mathbf{R}$. In Example 3.1.9 the component functions are

$$F_1(x, y, z) = x - y,\ F_2(x, y, z) = y^2,\ F_3(x, y, z) = y + z^2,\ F_4(x, y, z) = x + z.$$

Notice that $F_i : \mathbf{R}^3 \to \mathbf{R}$, $i = 1, 2, 3, 4$.

The function F may be defined only on a proper subset U of $\mathbf{R}^q$. Unless otherwise stated we assume that U is open and we write $F : U \to \mathbf{R}^p$.

Example 3.1.10 Let $F(x, y) = (x/(x - y), x + y)$. We see that F is defined only on the proper subset $U \subset \mathbf{R}^2$, $U = \{ (x, y) : x - y \neq 0 \}$.

A function $F: \mathbf{R}^q \to \mathbf{R}^p$ is said to be **continuous at a point** $\mathbf{x} \in \mathbf{R}^q$ if for every ball B_2 centered at $F(\mathbf{x})$ there exists a ball B_1 centered at $\mathbf{x}$ such that the image of every element of B_1 is in B_2.

An **equivalent definition** of continuity, which makes use of sequences of vectors, is the following.

First we say that a sequence of vectors $\{\mathbf{x}_n : n = 1, 2, ...\}$ converges to a vector $\mathbf{x}_0$ if the numerical sequence $\{ \|\mathbf{x}_n - \mathbf{x}_0\| : n = 1, 2, ...\}$ converges to 0. The continuity at $\mathbf{x}$ is defined by requiring that for every sequence $\{\mathbf{x}_n : n = 1, 2, ...\}$ which converges to $\mathbf{x}$, the corresponding sequence $\{F(\mathbf{x}_n) : n = 1, 2, ...\}$ converges to $F(\mathbf{x})$.

Notice that the sequence $\{F(\mathbf{x}_n)\}$ converges to $F(\mathbf{x})$ if and only if for each component F_i of F, the sequence of numbers $\{F_i(\mathbf{x}_n)\}$ converges to $F_i(\mathbf{x})$. Hence, a function F is continuous at $\mathbf{x}$ if and only if all its component functions are continuous at the same point $\mathbf{x}$.

A function $F: \mathbf{R}^q \to \mathbf{R}^p$ is said to be **continuous** if F is continuous at every $\mathbf{x}$. Checking the continuity of a function is sometimes a difficult task. We mention here a few helpful principles:

- *a sum or a difference of continuous functions is continuous ;*
- *a composition of continuous functions is continuous ;*
- *a polynomial function is continuous .*

Example 3.1.11 Let $F(x, y, z) = (x^2 - y, \text{Cos}\, xz)$. F is continuous, since its first component is simply a polynomial of degree 2 in the two variables x and y, and its second component is the composition of a polynomial of degree two with the continuous trigonometric function Cosine.

Notice that the norm function $\|.\| : \mathbf{R}^q \to \mathbf{R}$, which assigns to every vector $\mathbf{x}$ of $\mathbf{R}^q$ its norm, is continuous since, by Property iii) of the norm, we have

$$| \|\mathbf{x}_n\| - \|\mathbf{x}\| | \leq \|\mathbf{x}_n - \mathbf{x}\| .$$

This implies $\|\mathbf{x}_n\| \to \|\mathbf{x}\|$ when $\|\mathbf{x}_n - \mathbf{x}\| \to 0$.

Section 2. Matrices

1. Definitions and operations

A real square matrix, M, of size q is an array of real numbers with q rows and q columns. The element of the i-th row and j-th column is denoted by m_{ij}. We frequently write $M = \{m_{ij}, q\}$. Thus $\{m_{ij}, 3\}$ is a square matrix of size 3.

Example 3.2.1 In this example the first matrix is of size 2 and the second is of size 3. The entry of row 3 and column 1 is $m_{31} = 2$ (read *m three one*, not *m thirty-one*).

$$\begin{pmatrix} 1 & 2 \\ 2 & 3 \end{pmatrix}, \quad \begin{pmatrix} 1 & 2 & 3 \\ 1 & 2 & 1 \\ 2 & 3 & 2 \end{pmatrix}.$$

Each row or column of a square matrix of size q can be considered as a vector of $\mathbf{R}^q$. The matrix of size q whose entries are all 0 is called the **zero** matrix, and it is denoted by O_q, or simply O.

A matrix of size q whose entries satisfy the equalities $m_{ij} = 0$ if $i \neq j$, is called a **diagonal** matrix. The **identity** matrix of size q is a diagonal matrix in which $m_{ii} = 1$ for all $i = 1, 2, ..., q$. It is denoted by I_q, or simply I. Two matrices of the same size are **equal** if all corresponding entries are equal. A matrix M is **upper (lower) triangular** if $m_{ij} = 0$ whenever $j < i$ $(i < j)$.

Example 3.2.2

$$\begin{pmatrix} 1 & 0 & 0 \\ 0 & 1 & 0 \\ 0 & 0 & 1 \end{pmatrix} = I_3, \quad \begin{pmatrix} 1 & 2 & 3 \\ 0 & 3 & 1 \\ 0 & 0 & 2 \end{pmatrix}: \text{upper triangular}, \quad \begin{pmatrix} 1 & 0 & 0 \\ 2 & 3 & 0 \\ 4 & 3 & 2 \end{pmatrix}: \text{lower triangular}.$$

A matrix M is **symmetric** if $m_{ij} = m_{ji}$ for all i and j. The **transpose** M^T of a matrix M is obtained from M by replacing rows 1, 2, ..., q with columns 1, 2, ..., q.

Example 3.2.3

$$M = \begin{pmatrix} 1 & 5 \\ 3 & 4 \end{pmatrix}, \qquad M^T = \begin{pmatrix} 1 & 3 \\ 5 & 4 \end{pmatrix}.$$

Notice that a matrix is symmetric if and only if $M = M^T$.

A matrix M is **multiplied** by a number r as follows

a) $rM = \{ r\, m_{ij}, q \}$.

Matrices of the same size are **added** according to the rule

b) $M + P = \{ m_{ij}, q \} + \{ p_{ij}, q \} = \{ m_{ij} + p_{ij}, q \}$.

They are **multiplied row by column** as follows

c) $MP = S = \{ s_{ij}, q \}$,

where the element s_{ij} is obtained by taking the inner product of the i-th row of M and the j-th column of P (recall that both can be considered as vectors of $\mathbf{R}^q$). Thus

$$s_{ij} = a_{i1}p_{1j} + a_{i2}p_{2j} + \ldots + a_{iq}p_{qj} .$$

Example 3.2.4 Let

$$M = \begin{pmatrix} 1 & 5 \\ 3 & 4 \end{pmatrix}, \quad P = \begin{pmatrix} 2 & 3 \\ 1 & 6 \end{pmatrix} \text{ and } r = -2.$$

Then

$$rM = \begin{pmatrix} -2 & -10 \\ -6 & -8 \end{pmatrix}, \qquad M + P = \begin{pmatrix} 3 & 8 \\ 4 & 10 \end{pmatrix},$$

$$MP = \begin{pmatrix} 7 & 33 \\ 10 & 33 \end{pmatrix}, \qquad PM = \begin{pmatrix} 11 & 22 \\ 19 & 29 \end{pmatrix}.$$

Matrices of the same size satisfy the properties

1. $M + P = P + M$;
2. $(M + P) + R = M + (P + R)$;
3. $M(PR) = (MP)R$;
4. $M(P + R) = MP + MR$;
   $(P + R)M = PM + RM$.

The first is the commutative property of the sum. The second and the third are respectively the associative property of the sum and of the product. The fourth is the distributive property of the product with respect to the sum.

Notice that the product of matrices is not commutative, i.e. the property MP = PM does not hold (see Example 3.2.4).

2. **Determinant and rank**

To define the determinant of a matrix we first need the ideas of permutation and parity.

Given q distinct objects $O_1, O_2, ..., O_q$, consider the set A of all possible ordered arrangements, called **permutations** of size q, of the q objects. In every permutation each object appears exactly once. Two arrangements differ by the order in which the q objects are listed.

There are $q \times (q-1) \times ... \times 2 \times 1 = q!$ (read q factorial) elements in A, since the first entry of a permutation of size q can be selected in q ways, the second in q-1 ways, ... and the last in only one way.

Example 3.2.5 Let q = 5. Then 12345 (read: one, two, three, four, five), 31245, 54312 are permutations of the five distinct digits 1, 2, 3, 4, 5.

A permutation is said to be **even** (**odd**) if the natural increasing order $O_1, O_2, ..., O_q$ can be restored with an even (odd) number of switches between pairs.

Example 3.2.6 The permutation 31245 is even since the natural order can be restored with two switches.

Exchange 1 and 3 : 31245 becomes 13245 .
Exchange 2 and 3 : 13245 becomes 12345 .

The permutation 54312 is odd.

Exchange 5 and 1 : 54312 becomes 14352.
Exchange 4 and 2 : 14352 becomes 12354.
Exchange 4 and 5 : 12354 becomes 12345 .

Any permutation of size q can be identified with a one-to-one and onto function σ from the set of integers $N_q = \{1, 2, ..., q\}$ into itself. For example the permutation 54312 can be identified with the function $\sigma : N_5 \to N_5$,

$$\sigma(1) = 5\ , \sigma(2) = 4, \sigma(3) = 3, \sigma(4) = 1, \sigma(5) = 2.$$

The **parity** of a permutation, or the parity of the function σ which represents it, is 0 if the permutation is even and 1 if the permutation is odd.

For each matrix M of size q we define the **determinant** of M, denoted by $\det M$ or $|M|$, as the number

$$\det M = \sum_{\sigma}(-1)^{\text{par}\sigma}a_{1\sigma(1)}a_{2\sigma(2)}\ldots a_{q\sigma(q)} \tag{3.2.1}$$

where σ ranges over all $q!$ permutations of $1, 2, ..., q$ and $\text{par}\sigma$ is the parity of the permutation. Notice that the first index of a_{ij} follows the natural increasing order of 1, 2, ..., q and only the second index is affected by the permutation. Formula (3.2.1) tells us that the determinant of M is obtained by first taking all possible distinct products among the entries of M, with each product containing exactly one element from each row and each column. The elements of every product are listed in increasing row-order. The positions of the elements in the columns of M form a permutation which decides the sign of the product. If the permutation is of odd parity the sign of the corresponding product must be changed. All numbers so obtained are then added to give $\det M$.

Example 3.2.7 With

$$M = \begin{pmatrix} a_{11} & a_{12} \\ a_{21} & a_{22} \end{pmatrix},$$

we have $\det M = a_{11}a_{22} - a_{12}a_{21}$.

With

$$N = \begin{pmatrix} a_{11} & a_{12} & a_{13} \\ a_{21} & a_{22} & a_{23} \\ a_{31} & a_{32} & a_{33} \end{pmatrix},$$

we have $\det N = a_{11}a_{22}a_{33} + a_{12}a_{23}a_{31} + a_{13}a_{21}a_{32} - a_{11}a_{23}a_{32} - a_{12}a_{21}a_{33} - a_{13}a_{22}a_{31}$.

The determinant has the following basic properties, which easily follow from its definition.

a) $\det I = 1$.

b) The determinant of the matrix P obtained from M by exchanging either two rows or two columns of M has the property $\det P = -\det M$.

c) The determinant of the matrix Q obtained from M by adding to any row (or any column) of M a multiple of any row (or column) is the same as the determinant of M.

From formula (3.2.1) we see that the determinant of a matrix having one zero row or column is zero, while from property b) we derive that the determinant of a matrix with two equal rows or columns is also zero.
More generally, we have the following result.

Proposition 3.2.1 *The determinant of a matrix* M *is zero if and only if the row vectors (or column vectors) of* M *are linearly dependent (one row or one column is a linear combination of the other rows or columns).*

Three useful additional properties of determinants are,

d) $\det MP = \det M \det P$;

e) $\det rI_q = r^q$;

f) $\det M = \det M^T$.

The computation of the determinant of a matrix M is today made easier by several computer packages. For matrices of size 2, 3 this task is easily accomplished by hand, as we have seen in Example 3.2.7.

There is also a recursion formula which can be used for computing the determinant. It states that given any row index $i = 1, 2, ..., q$, we have

$$(3.2.2) \qquad \det M = \sum_{j=1}^{q} (-1)^{i+j} a_{ij} \det M_{ij}$$

where M_{ij} is the matrix of size $q-1$ obtained from M by deleting row i and column j. Since the determinant of a matrix of size 1, $M = \{a_{11}\}$, is obviously a_{11}, we see that formula (3.2.2) allows the computation of determinants by recursion.

It can be shown that (3.2.2) gives detM regardless of the row i used in the computation. Moreover, from property f) we derive that in (3.2.2) the roles of rows and columns can be exchanged. For example, the determinant of the matrix $N = \{a_{ij}, 3\}$ can be computed using the first row or the first column.
We obtain

$$\det N = a_{11}(a_{22}a_{33} - a_{23}a_{32}) - a_{12}(a_{21}a_{33} - a_{23}a_{31}) + a_{13}(a_{21}a_{32} - a_{22}a_{31}), \text{ using row 1;}$$

$$\det N = a_{11}(a_{22}a_{33} - a_{23}a_{32}) - a_{21}(a_{12}a_{33} - a_{13}a_{32}) + a_{31}(a_{12}a_{23} - a_{13}a_{22}), \text{ using column 1.}$$

Both results coincide with the value obtained in Example 3.2.7.

Example 3.2.8 Compute the determinant of the matrix

$$M = \begin{pmatrix} 1 & 3 & 2 \\ 0 & 2 & 1 \\ 2 & 3 & 1 \end{pmatrix}$$

using the first row and the second column.

Using the first row we have

$$\det M = 1(2 - 3) - 3(0 - 2) + 2(0 - 4) = -1 + 6 - 8 = -3 .$$

Using the second column we have

$$\det M = -3(0 - 2) + 2(1 - 4) - 3(1 - 0) = 6 - 6 - 3 = -3.$$

Let M be a matrix of size q and let $p < q$ be a positive integer. Any matrix derived from M by deleting p rows and p columns of M is called a **minor** of M. The size of the minor is $q - p$. For $p = 0$, the only minor is M itself, while for $p = q - 1$ the q^2 minors are simply the entries of M.

The **rank** of a matrix M, denoted by rankM, is the size of the largest minor of M with determinant different from 0. Thus, rankM = q if and only if $\det M \neq 0$. The only matrix with zero rank is the zero matrix.

Example 3.2.9 The matrix M of Example 3.2.8 has rank 3 . The matrix

$$N = \begin{pmatrix} 0 & 0 & 0 \\ 0 & 1 & 1 \\ 0 & 1 & 1 \end{pmatrix}$$

has rank 1.

3. Invertible matrices and similarity

Let M be a matrix of size q with determinant different from 0. We can produce a matrix P of the same size, such that

(3.2.3) $$MP = PM = I.$$

The matrix $P = \{p_{ij}, q\}$ is defined as follows

(3.2.4) $$p_{ij} = (-1)^{i+j} \frac{\det M_{ji}}{\det M},$$

where M_{ji} is the matrix of size q-1 obtained from M by deleting row j and column i. Notice the exchange of indices between p_{ij} and M_{ji}.

Example 3.2.10 Let M be the matrix

$$M = \begin{pmatrix} 1 & 0 & 1 \\ 2 & 1 & 0 \\ 0 & 1 & 2 \end{pmatrix}.$$

Then
$\det M = 4$, $\det M_{11} = 2$, $\det M_{12} = 4$, $\det M_{13} = 2$, $\det M_{21} = -1$, $\det M_{22} = 2$, $\det M_{23} = 1$, $\det M_{31} = -1$, $\det M_{32} = -2$, $\det M_{33} = 1$.
Consequently ,

$$P = \begin{pmatrix} 0.5 & 0.25 & -0.25 \\ -1 & 0.5 & 0.5 \\ 0.5 & -0.25 & 0.25 \end{pmatrix}.$$

In general, the construction of P is done with numerical packages, at least whenever the size of the matrix is large (≥ 4).

Let us check that MP = I. Recall that the element s_{ik} of MP is obtained by taking the inner product of row i of M and column k of P. Thus,

$$s_{ik} = \sum_{j=1}^{q} a_{ij}\, p_{jk} = \frac{1}{\det M}\sum_{j=1}^{q} (-1)^{j+k} a_{ij} \det M_{kj} .$$

We see that for $k = i$ the summation on the right gives the determinant of M, according to formula (3.2.2). Hence $s_{ii} = 1$ for all $i = 1, 2, ..., q$. For $k \neq i$ the summation is zero since it represents the determinant of the matrix K obtained from M by replacing row k with row i. Thus K has two equal rows and $\det K = 0$. The equality $PM = I$ can be checked in an analogous manner

A matrix M of size q such that the equation $MX = I$ is satisfied for some matrix X of the same size is said to be **invertible** and X is called the **inverse** of M. We have the following important result.

Theorem 3.2.1 M *is invertible if and only if* $\det M \neq 0$. *Moreover the inverse of* M *is unique.*

Proof. We have seen before that if $\det M \neq 0$ we can construct a matrix P such that $MP = PM = I$. Hence M is invertible. Vice versa, if M is invertible, then we can find X such that $MX = I$. Since $\det MX = \det M \det X = \det I = 1$, we must have $\det M \neq 0$.

Assume now that M is invertible. We know that $\det M \neq 0$ and we can construct P such that $PM = MP = I$. Let $MX = I$ for some matrix X. Then $X = (PM)X = P(MX) = P$. Hence $X = P$ and the inverse of M is unique.

We usually denote the inverse of an invertible matrix M with the symbol M^{-1}. Obviously $(M^{-1})^{-1} = M$.

Notice that the invertibility of M and of Q implies the invertibility of MQ, since $\det MQ = \det M \det Q$. Moreover,

(3.2.5) $$(MQ)^{-1} = Q^{-1}M^{-1}$$

since $(MQ)(Q^{-1}M^{-1}) = M(QQ^{-1})M^{-1} = MM^{-1} = I$.
In particular, with $Q = M$ we obtain $(M^2)^{-1} = (M^{-1})^2$ and, by means of an inductive argument,

(3.2.6) $$(M^n)^{-1} = (M^{-1})^n$$

for every $n \geq 1$.

Let M and N be two matrices of the same size. We say that M and P are **similar** if there exists an invertible matrix T such that

$$(3.2.7) \qquad N = T^{-1}MT.$$

Notice that $\det N = \det M$ since $\det T^{-1} \det T = \det TT^{-1} = \det I = 1$. Moreover, we see from (3.2.7) that N and M are similar if and only if $TN = MT$ for some invertible matrix T.

Example 3.2.11 Let

$$M = \begin{pmatrix} 0.5 & 2 \\ 0 & 0 \end{pmatrix}, \qquad N = \begin{pmatrix} 0.5 & 0 \\ 0 & 0 \end{pmatrix}.$$

Then $TN = MT$, where T is the matrix $T = \begin{pmatrix} 1 & 4 \\ 0 & -1 \end{pmatrix}$.

We shall see later how to check if two matrices are similar and how to construct T.

4. Eigenvalues and spectral radius

Let M be a matrix of size q, and consider the q-th degree polynomial, called the **characteristic polynomial** of M,

$$(3.2.8) \qquad P(r) = \det(rI - M).$$

Notice that the matrix $rI - M$ is invertible if and only if $P(r) \neq 0$. The q-th degree equation $P(r) = 0$ may have no real roots. However, by the Fundamental Theorem of Algebra, there exist $p \leq q$ numbers, $r_1, r_2, \ldots, r_p$, real or complex, and p positive integers $m_1, m_2, \ldots, m_p$ such that $m_1 + m_2 + \ldots + m_p = q$ and

$$(3.2.9) \qquad P(r) = (r - r_1)^{m_1} (r - r_2)^{m_2} \ldots (r - r_p)^{m_p}.$$

Every root of (3.2.9) is called an **eigenvalue** of the matrix M. Moreover, if for some $k = 1, 2, \ldots, p$, the root r_k is a complex number, namely $r_k = a + \mathbf{i}b$ with $b \neq 0$ (recall that $\mathbf{i}^2 = -1$), then the complex conjugate number $a - \mathbf{i}b$ is also a root of P(r) and the term $(r - (a - \mathbf{i}b))$ appears in (3.2.9) with the same exponent of $(r - (a + \mathbf{i}b))$.

The set of all eigenvalues of M (real or complex) is called the **spectrum of M**. We shall denote it by SpecM.

Example 3.2.12 Let

$$M = \begin{pmatrix} 0 & -1 \\ 1 & 0 \end{pmatrix} \text{ and } Q = \begin{pmatrix} 1 & 2 \\ 2 & 3 \end{pmatrix}.$$

Then

$$\det(rI - M) = r^2 + 1 \text{ and } \det(rI - Q) = (r - 1)(r - 3) - 4 = r^2 - 4r - 1.$$

The equation $r^2 + 1 = 0$ does not have any real root. In fact

$$r^2 + 1 = (r - \mathbf{i})(r + \mathbf{i}) .$$

The equation $r^2 - 4r - 1 = 0$ has two real roots, namely $2 \pm \sqrt{5}$. Therefore

$$r^2 - 4r - 1 = (r - 2 - \sqrt{5})\,(r - 2 + \sqrt{5}).$$

The spectrum of M is $\{\mathbf{i}, -\mathbf{i}\}$ and the spectrum of N is $\{2 + \sqrt{5}, 2 - \sqrt{5}\}$.

For similar matrices we have the following result.

Theorem 3.2.2 *Let* N *and* M *be two similar matrices. Then* SpecM = SpecN .

Proof. Since N and M are similar matrices, there exists an invertible matrix T such that $N = T^{-1}MT$. Then $rI - N = T^{-1}rIT - T^{-1}MT = T^{-1}(rI - M)T$. Thus $\det(rI - N) = \det(rI - M)$ and the result follows.

Matrices with equal spectrum may not be similar as the following example shows.

Example 3.2.13 Let

$$M = \begin{pmatrix} 1 & 0 \\ 0 & 1 \end{pmatrix}, \ N = \begin{pmatrix} 1 & 1 \\ 0 & 1 \end{pmatrix} \text{ and } T = \begin{pmatrix} a & b \\ c & d \end{pmatrix}.$$

Notice that SpecM = SpecN = {1}. From TN = MT we derive TN = T, which implies a = c = 0. Hence T cannot be invertible.

Recall that the modulus $|z|$ of a complex number $z = a + ib$ is $|z| = \sqrt{a^2 + b^2}$.

The **spectral radius of a matrix M**, denoted by $\rho(M)$, is the modulus of its largest eigenvalue, i.e.

$$\rho(M) = \max\{|r| : r \in \mathrm{Spec}M\}. \tag{3.2.10}$$

The matrix $rI - M$ is invertible whenever $|r| > \rho(M)$.
The following result will be useful later.

Theorem 3.2.3 *Let* M *be an invertible matrix. Then, the eigenvalues of* M^{-1} *are the reciprocals of the eigenvalues of* M.

<u>Proof</u>. Let r be an eigenvalue of M, i.e. $\det(rI - M) = 0$. Since $\det M \neq 0$ we obtain that $r \neq 0$. Let $s = r^{-1}$. We want to show that $\det(sI - M^{-1}) = 0$.
Suppose not. Then $\det(sI - M^{-1})M = \det(sI - M^{-1})\det M \neq 0$ and

$$(sI - M^{-1})M = sM - I$$

is invertible. This result is impossible since

$$sM - I = -I + sM = (-sI)(rI - M) \text{ and } \det(rI - M) = 0.$$

Hence $\det(sI - M^{-1}) = 0$ and $s \in \mathrm{Spec}M^{-1}$. Since the argument is symmetric in r and s, the result follows.

The spectral radius of M^{-1} is the modulus of the reciprocal of the smallest eigenvalue of M.

Example 3.2.14 Let

$$M = \begin{pmatrix} 1 & 2 \\ 4 & 3 \end{pmatrix}.$$

Then

$$M^{-1} = -\frac{1}{5}\begin{pmatrix} 3 & -2 \\ -4 & 1 \end{pmatrix},$$

and

$$\mathrm{Spec}M = \{5, -1\},\ \mathrm{Spec}M^{-1} = \{0.2, -1\},\ \rho(M) = 5,\ \rho(M^{-1}) = 1.$$

Section 3. Matrices as linear operators

1. The action of a matrix

Let M be a matrix of size q. We define an action of M on $\mathbf{R}^q$ by setting

(3.3.1) $$M\mathbf{x} = (\mathbf{m}_1 \cdot \mathbf{x}, \mathbf{m}_2 \cdot \mathbf{x}, ..., \mathbf{m}_q \cdot \mathbf{x})$$

where $\mathbf{m}_j$, i = 1, 2, ..., q, represents the j-th row of M.

Example 3.3.1 Let M be the matrix of size 2,

$$M = \begin{pmatrix} 2 & 3 \\ 1 & 5 \end{pmatrix}$$

Then $M\mathbf{x} = ((2, 3) \cdot \mathbf{x}, (1, 5) \cdot \mathbf{x}) = (2x + 3y, x + 5y)$.

The following properties of the action of M are important.

1) For every pair of real numbers a, b and for every pair of vectors **x, y** we have $M(a\mathbf{x} + b\mathbf{y}) = aM\mathbf{x} + bM\mathbf{y}$.
2) For every pair of matrices M, P of size q, and for every vector **x**, we have $P(M\mathbf{x}) = (PM)\mathbf{x}$, where PM is the row by column product previously defined.

Property 1 is described by saying that the action of M, as defined in (3.3.1), is **linear**. Property 2 tells us that the action of M followed by the action of P has the same effect as the action of the matrix PM.

The justification of Property 1 is straightforward and it is based on the properties of the inner product. Property 2 is proved by writing the coordinates of $P(M\mathbf{x})$, $(PM)\mathbf{x}$ and by verifying that they are equal.

In the previous section we have given a **static** definition of equality between matrices. We can give a **dynamic** definition of equality between matrices by saying that two matrices M and P of the same size q are **dynamically equal** if for every $\mathbf{x} \in \mathbf{R}^q$ we have $M\mathbf{x} = P\mathbf{x}$. The following proposition links the two definitions.

Proposition 3.3.1 *Two matrices* M *and* P *are dynamically equal if and only if they are statically equal.*

Proof. If $M = P$ in the static sense, then obviously $M = P$ in the dynamic sense. Assume that $M = P$ in the dynamic sense. Then $Me_i = Pe_i$. Since

$$Me_i = (m_{1i}, m_{2i}, ..., m_{qi}) \text{ , and similarly } Pe_i = (p_{1i}, p_{2i}, ..., p_{qi}) ,$$

we see that corresponding columns of M and P are equal. Hence M and P are statically equal.

Remark 3.3.1 Notice that M and P are dynamically equal if and only if $Me_i = Pe_i$ for every $\mathbf{e}_i, i = 1, 2, ..., q$. In fact from

$$\mathbf{x} = a_1\mathbf{e}_1 + ... + a_q\mathbf{e}_q \text{ we obtain } M\mathbf{x} = a_1Me_1 + ... + a_qMe_q, P\mathbf{x} = a_1Pe_1 + ... + a_qPe_q.$$

Thus, $M\mathbf{x} = P\mathbf{x}$ for every $\mathbf{x} \in \mathbf{R}^q$ if and only if $Me_i = Pe_i$ for every $\mathbf{e}_i, i = 1, 2, ..., q$.

This remark tells us that knowing the action of M on the vectors of the standard basis amounts to know M entirely.

Example 3.3.2 Let M be a square matrix such that $Me_1 = (3, -2), Me_2 = (-1, 5)$. Then, for every $\mathbf{x} \in \mathbf{R}^2$ we have $M\mathbf{x} = (3x - y, - 2x + 5y)$. In fact,

$$\mathbf{x} = x\mathbf{e}_1 + y\mathbf{e}_2 \text{ and } M\mathbf{x} = xMe_1 + yMe_2$$

imply

$$M\mathbf{x} = x(3, -2) + y(-1, 5) = (3x - y, - 2x + 5y).$$

Since $(3x - y, - 2x + 5y) = ((3, -1)\bullet\mathbf{x}, (-2, 5)\bullet\mathbf{x})$ we obtain

$$M = \begin{pmatrix} 3 & -1 \\ -2 & 5 \end{pmatrix}.$$

2. **Operator norm of a matrix**

From the Cauchy-Schwarz Inequality, we derive

$$||M\mathbf{x}||^2 = (\mathbf{m}_1\bullet\mathbf{x})^2 + ... + (\mathbf{m}_q\bullet\mathbf{x})^2 \leq (||\mathbf{m}_1||^2 + ... + ||\mathbf{m}_q||^2) ||\mathbf{x}||^2 = p^2||\mathbf{x}||^2 .$$

with $p^2 = ||\mathbf{m}_1||^2 + ... + ||\mathbf{m}_q||^2$. Hence,

$$||M\mathbf{x} - M\mathbf{y}|| = ||M(\mathbf{x} - \mathbf{y})|| \leq p \, ||\mathbf{x} - \mathbf{y}||.$$

This inequality implies the continuity of M.
Define

$$(3.3.2) \qquad \|M\| = \sup\{ \|M\mathbf{x}\| : \|\mathbf{x}\| = 1 \} \le \sup\{ p\|\mathbf{x}\| : \|\mathbf{x}\| = 1 \} = p.$$

The number $\|M\|$ is called the **operator norm** of M induced by the Euclidean norm in $\mathbf{R}^q$. For every vector $\mathbf{y} \in \mathbf{R}^q$ we have

$$(3.3.3) \qquad \|M\mathbf{y}\| \le \|M\| \, \|\mathbf{y}\| .$$

The inequality is obvious if $\mathbf{y} = \mathbf{0}$ since $M\mathbf{0} = \mathbf{0}$. For every other vector $\mathbf{y}$ consider the vector $\mathbf{x} = \mathbf{y}/\|\mathbf{y}\|$. Then $\|\mathbf{x}\| = 1$ and, according to (3.3.2), $\|M\mathbf{x}\| \le \|M\|$. Inequality (3.3.3) follows from property ii) of the norm, which implies $\|M\mathbf{x}\| = (1/\|\mathbf{y}\|)\|M\mathbf{y}\|$.

Let M and P be two matrices of size q and let S = PM. Then the following basic inequality is easily established from the definition of the operator norm.

$$(3.3.4) \qquad \|S\| \le \|P\| \, \|M\| .$$

In particular, for P = M we obtain $\|M^2\| \le \|M\|^2$, and in general

$$(3.3.5) \qquad \|M^n\| \le \|M\|^n .$$

Consequently, the sequence $\{\sqrt[n]{\|M^n\|} , n = 1, 2, ...\}$ is bounded above by $\|M\|$. Equality (3.3.6) below tells us that the sequence is in fact convergent and its limit is the spectral radius of M:

$$(3.3.6) \qquad \rho(M) = \lim_{n\to\infty} \sqrt[n]{\|M^n\|} .$$

Equality (3.3.6) holds true even if a different norm $\|.\|_N$ is considered in $\mathbf{R}^q$ such as the taxi-cab norm, and the operator norm of M^n, n = 1, 2, ... is defined using the new norm, i.e. $\|M^n\|_N = \sup\{\|M^n\mathbf{x}\|_N : \|\mathbf{x}\|_N = 1\}$. In this way the numerical value of the operator norm of M^n usually changes, but equality (3.3.6) remains valid.

Example 3.3.3 Let

$$M = \begin{pmatrix} 1 & 0 \\ 1 & 2 \end{pmatrix}.$$

Then $\rho(M) = 2$, since SpecM = $\{1, 2\}$.

Moreover

$$M^2 = \begin{pmatrix} 1 & 0 \\ 1+2 & 2^2 \end{pmatrix}, \quad M^3 = \begin{pmatrix} 1 & 0 \\ 1+2+2^2 & 2^3 \end{pmatrix},$$

and in general,

$$M^n = \begin{pmatrix} 1 & 0 \\ 2^n - 1 & 2^n \end{pmatrix},$$

since $1 + 2 + 2^2 + \ldots + 2^{n-1} = 2^n - 1$. Let (x, y) be such that $x^2 + y^2 = 1$. From

$$\|M^n(x, y)\|^2 = 2^{2n} + 2x(2^n - 1)(2^n y - x),$$

we derive

$$2^n\sqrt{2} \geq \|M^n\| = \sup\{\|M^n\mathbf{x}\| : \|\mathbf{x}\| = 1\} \geq \|M^n(0,1)\| = 2^n.$$

Thus

$$\lim_{n\to\infty} \sqrt[n]{\|M^n\|} = 2$$

as expected, since $\rho(M) = 2$.

In Chapter 6 we present a numerical method for estimating the spectral radius of M without having to find the spectrum of M or the norm of M^n.

3. Kernel, image and generalized kernel

The set of vectors $\mathbf{x}$ of $\mathbf{R}^q$ such that $M\mathbf{x} = \mathbf{0}$ is called **kernel** of M and it is denoted by KerM. The set of vectors $\mathbf{y}$ of $\mathbf{R}^q$ such that $\mathbf{y} = M\mathbf{z}$ for some $\mathbf{z} \in \mathbf{R}^q$ is called **image** of M and it is denoted by ImM.
KerM and ImM are subspaces of $\mathbf{R}^q$.

In fact, let $\mathbf{x}, \mathbf{y} \in$ KerM, i.e. $M\mathbf{x} = M\mathbf{y} = \mathbf{0}$, and let $\mathbf{z} = a\mathbf{x} + b\mathbf{y}$. Then,

$$M\mathbf{z} = M(a\mathbf{x} + b\mathbf{y}) = aM\mathbf{x} + bM\mathbf{y} = a\mathbf{0} + b\mathbf{0} = \mathbf{0}$$

which implies $\mathbf{z} \in$ KerM.
Similarly, let $\mathbf{u}, \mathbf{v} \in$ ImM, i.e. $\mathbf{u} = M\mathbf{x}$, $\mathbf{v} = M\mathbf{y}$, for some $\mathbf{x}, \mathbf{y} \in \mathbf{R}^q$ and let $\mathbf{z} = a\mathbf{u} + b\mathbf{v}$.

We have

$$\mathbf{z} = aM\mathbf{x} + bM\mathbf{y} = M(a\mathbf{x} + b\mathbf{y}).$$

This implies $\mathbf{z} \in \mathrm{Im}M$.

Example 3.3.4 Let

$$M = \begin{pmatrix} 1 & 2 \\ 2 & 4 \end{pmatrix}.$$

From $M\mathbf{x} = M(x, y) = (x + 2y, 2x + 4y)$ we derive $\mathrm{Ker}M = \{(x, y) : x + 2y = 2x + 4y = 0\}$. Hence $\mathbf{x} \in \mathrm{Ker}M$ if and only if $x + 2y = 0$. This equation represents a line passing through the origin, and is therefore a one-dimensional subspace of $\mathbf{R}^2$.

A point $(a, b) \in \mathrm{Im}M$ if and only if there exists (x, y) such that $x + 2y = a$ and $2x + 4y = b$. Thus $2a = 2x + 4y = b$. The image of M is made of those vectors (a, b) such that $b = 2a$, again a line passing through the origin.

In Example 3.3.4 we have $\dim \mathrm{Im}M + \dim \mathrm{Ker}M = 2$. In general, it can be shown that for every matrix M of size q one has the important equality

(3.3.7) $$\dim \mathrm{Im}M + \dim \mathrm{Ker}M = q .$$

Notice that $\mathrm{Ker}M \subset \mathrm{Ker}M^2$. Therefore $\dim \mathrm{Ker}M^2 \geq \dim\mathrm{Ker}M$.

Example 3.3.5 Let

$$M = \begin{pmatrix} 0 & 1 \\ 0 & 0 \end{pmatrix}.$$

The matrix M^2 is the zero matrix . Hence $\mathrm{Ker}M^2 = \mathbf{R}^2$, while $\mathrm{Ker}M = \{(x, y) : y = 0\}$.

Similarly, we have $\mathrm{Ker}M^p \subset \mathrm{Ker}M^{p+1}$ and

$$\dim\mathrm{Ker}M^{p+1} \geq \dim\mathrm{Ker}M^p.$$

Since $M^q\mathbf{x} \neq \mathbf{0}$ implies $M^{q+i}\mathbf{x} \neq \mathbf{0}$ for all $i \geq 1$, there exists $p \leq q$ such that

(3.3.8) $$\mathrm{Ker}M^{p+1} = \mathrm{Ker}M^p .$$

The subspace $\mathrm{Ker}M^p$ is called the **generalized kernel** of M .

We shall see in the following section that this concept is extremely useful. Since $\mathrm{Ker}M^{q+1} = \mathrm{Ker}M^q$, the kernel of M^q always coincides with the generalized kernel of M.

4. Eigenvectors

Let $\mathbf{x} \neq \mathbf{0}$ be a vector of $\mathbf{R}^q$ and let k be a number such that

$$(3.3.9) \qquad M\mathbf{x} = k\mathbf{x}.$$

Then $\mathbf{x}$ is called an **eigenvector** of the matrix M corresponding to k. Since (3.3.9) can be rewritten as

$$(3.3.10) \qquad (kI - M)\mathbf{x} = \mathbf{0},$$

we see that kI - M cannot be invertible for otherwise, by applying $(kI-M)^{-1}$ to both sides of (3.3.10), we would obtain

$$(3.3.11) \qquad \mathbf{x} = (kI - M)^{-1}\mathbf{0} = \mathbf{0}$$

against the assumption $\mathbf{x} \neq \mathbf{0}$. Hence, $\det(kI - M) = 0$ and k is an **eigenvalue** of the matrix M. Notice that for every $t \neq 0$, the vector $\mathbf{y} = t\mathbf{x}$ is also an eigenvector corresponding to the eigenvalue k since $M\mathbf{y} = Mt\mathbf{x} = tM\mathbf{x} = tk\mathbf{x} = k(t\mathbf{x}) = k\mathbf{y}$.

The number $gm = \dim\mathrm{Ker}(kI - M)$ is called **geometric multiplicity** of the eigenvalue k, while the dimension of the generalized kernel of kI - M is called **algebraic multiplicity** of the eigenvalue k. Denoting this number by am we obtain, from our previous remarks, $am = \dim\mathrm{Ker}(kI - M)^q$. Obviously $am \geq gm$.

The algebraic multiplicity of k coincides with the exponent of r - k in the characteristic polynomial of M, while its geometric multiplicity equals q - rank(kI - M). When the algebraic multiplicity of k is one, the geometric multiplicity is also one, and the eigenvalue is called **simple**.

Particularly important are those matrices M for which the geometric multiplicity of each eigenvalue is equal to its algebraic multiplicity. This is the case if, for example, M is symmetric. Such eigenvalues are called **semisimple**.

We have seen that whenever (3.3.9) is satisfied by some real number k and some vector $\mathbf{x} \neq \mathbf{0}$, then $k \in \mathrm{Spec}M$.

Vice versa, assume that $k \in \text{Spec}M$ and k is a real number. Then $\det(kI - M) = 0$ and according to Proposition 3.2.1, the columns of $kI - M$ are linearly dependent. Hence there are constants $x_1, x_2, ..., x_{q-1}, x_q$, not all zero, such that

$$(k - m_{11})x_1 - m_{12}x_2 - ... - m_{1q-1}x_{q-1} - m_{1q}x_q = 0,$$

$$..................................$$

$$-m_{q1}x_1 - m_{q2}x_2 - ... - m_{qq-1}x_{q-1} + (k - m_{qq})x_q = 0.$$

These equations coincide with the vector equation

$$(kI - M)\mathbf{x} = \mathbf{0} \text{ with } \mathbf{x} = (x_1, ..., x_q) \neq \mathbf{0}.$$

Thus $\mathbf{x}$ is an eigenvector of M.

For every real eigenvalue k of M, the subspace $\mathbf{X} = \text{Ker}(kI - M)$ is invariant under the action of M, i.e. $M\mathbf{x} \in \mathbf{X}$ for every $\mathbf{x} \in \mathbf{X}$. In fact, $\mathbf{x} \in \mathbf{X}$ implies $M\mathbf{x} = k\mathbf{x} \in \mathbf{X}$. We will prove later (Chapter 4) that the generalized kernel of $kI - M$ is also invariant under the action of M.

Example 3.3.6 Let

$$M = \begin{pmatrix} 1 & 0 & 1 \\ 0 & 2 & 0 \\ 0 & 1 & 2 \end{pmatrix}.$$

The eigenvalues of M are $k_1 = 1$ and $k_2 = k_3 = 2$. The eigenvalue 1 is simple. The eigenvalue 2 has algebraic multiplicity 2 and geometric multiplicity 1. In fact $\text{rank}(2I - M) = 2$. An eigenvector corresponding to 2 is $\mathbf{u} = (1, 0, 1)$. The subspace $\mathbf{X}$ spanned by $\mathbf{u}$ is invariant under M. More precisely, $M\mathbf{x} = 2\mathbf{x}$ for every $\mathbf{x} \in \mathbf{X}$. The generalized kernel $\mathbf{Y}$ of $2I - M$ has dimension 2. Since

$$(2I - M)^2 = \begin{pmatrix} 1 & 1 & -1 \\ 0 & 0 & 0 \\ 0 & 0 & 0 \end{pmatrix},$$

we find that $\mathbf{Y} = \text{Span}\{\mathbf{u}, \mathbf{v} = (0, 1, 1)\}$.

Let $\mathbf{x} = a\mathbf{u} + b\mathbf{v}$. Then $M\mathbf{x} = aM\mathbf{u} + bM\mathbf{v} = a(2, 0, 2) + b(1, 2, 3) = (2a + b)\mathbf{u} + 2b\mathbf{v}$. Therefore, $\mathbf{Y}$ is invariant under the action of M.

Example 3.3.7 Let

$$M = \begin{pmatrix} 3 & 0 & 1 \\ 0 & 4 & 0 \\ 0 & 0 & 4 \end{pmatrix}.$$

We have $\det(rI - M) = (r - 3)(r - 4)^2$. The eigenvalues of M are $r_1 = 3$ and $r_2 = 4$. Their algebraic multiplicity are 1 and 2 respectively. The geometric multiplicity of r_1 is 1. The geometric multiplicity of r_2 is 2. In fact

$$4I - M = \begin{pmatrix} 1 & 0 & -1 \\ 0 & 0 & 0 \\ 0 & 0 & 0 \end{pmatrix},$$

and $\text{rank}(4I - M) = 1$. Both eigenvalues are semisimple.

Two linearly independent eigenvectors corresponding to r_2 are found by solving the system

$$(4I - M)(x, y, z) = (0, 0, 0),$$

which implies $x - z = 0$. We can choose $\mathbf{u} = (1, 0, 1)$ and $\mathbf{v} = (0, 1, 0)$. The invariant subspace is the plane $x - z = 0$.

The matrix M can have complex conjugate eigenvalues . Assume that

$$r_1 = a + ib \qquad \text{and} \qquad r_2 = a - ib$$

are eigenvalues of M. It can be shown that there is a pair of linearly independent vectors $\mathbf{u}$ and $\mathbf{v}$ of $\mathbf{R}^q$ such that

$$M(\mathbf{u} + i\mathbf{v}) = (a + ib)(\mathbf{u} + i\mathbf{v}), \tag{3.3.12}$$

i. e.

$$M\mathbf{u} + iM\mathbf{v} = a\mathbf{u} - b\mathbf{v} + i(b\mathbf{u} + a\mathbf{v}). \tag{3.3.13}$$

Since the real and imaginary part of (3.3.13) must be equal we obtain

$$M\mathbf{u} = a\mathbf{u} - b\mathbf{v}, \qquad \text{and} \qquad M\mathbf{v} = b\mathbf{u} + a\mathbf{v}. \tag{3.3.14}$$

It follows from (3.3.14) that the two-dimensional subspace $\mathbf{X}$ spanned by the two linearly independent vectors $\mathbf{u}$ and $\mathbf{v}$ is invariant under the action of M.

In fact **Mu** and **Mv** are linear combinations of **u** and **v**, which implies

$$\mathbf{Mu} \in \mathbf{X} \text{ and } \mathbf{Mv} \in \mathbf{X}.$$

Consequently, for every $\mathbf{z} = \alpha\mathbf{u} + \beta\mathbf{v} \in \mathbf{X}$, we obtain

$$\mathbf{Mz} = \alpha\mathbf{Mu} + \beta\mathbf{Mv} = (\alpha a + \beta b)\mathbf{u} + (-\alpha b + \beta a)\mathbf{v} \in \mathbf{X}.$$

Thus, to the pair of complex conjugate eigenvalues $a + \mathbf{i}b$, $a - \mathbf{i}b$, there corresponds a two-dimensional invariant subspace of $\mathbf{R}^q$.

The simplest way to find **u** and **v** is to solve (3.3.12), following the guidelines of the next example.

Example 3.3.8 Let

$$M = \begin{pmatrix} 1 & 0 & 1 \\ 0 & 2 & 1 \\ 0 & -1 & 2 \end{pmatrix}.$$

The characteristic polynomial of M is $(r - 1)[(r - 2)^2 + 1]$ and the eigenvalues of M are the real number $r_1 = 1$, and the complex conjugate pair $r_2 = 2 + \mathbf{i}$, $r_3 = 2 - \mathbf{i}$. An eigenvector corresponding to 1 is found by solving the system

$$(I-M)\mathbf{x} = \begin{pmatrix} 0 & 0 & -1 \\ 0 & -1 & -1 \\ 0 & 1 & -1 \end{pmatrix}(x, y, z) = (0, 0, 0).$$

A solution is (1, 0, 0).

Equation (3.3.12) can be rewritten in the following form

$$(r_2I - M)\,\mathbf{w} = \begin{pmatrix} 1+\mathbf{i} & 0 & -1 \\ 0 & \mathbf{i} & -1 \\ 0 & 1 & \mathbf{i} \end{pmatrix}(x, y, z) = (0, 0, 0),$$

where $\mathbf{w} = (x, y, z) = \mathbf{x} + \mathbf{i}\mathbf{y}$. Notice that the coordinates (x, y, z) of $\mathbf{w}$ are complex numbers. We obtain

$$\begin{aligned} (1 + \mathbf{i})x \; - \; z &= 0, \\ \mathbf{i}y \; - \; z &= 0, \\ y + \mathbf{i}z &= 0. \end{aligned}$$

Since the second and third equation coincide (multiply the third by $\mathbf{i}$) we solve the first two. Setting $x = 1$, we get $z = 1 + \mathbf{i}$ and $\mathbf{i}y = 1 + \mathbf{i}$, which implies $y = 1 - \mathbf{i}$. Hence

$$\mathbf{w} = (1, 1 - \mathbf{i}, 1 + \mathbf{i}) = (1, 1, 1) + \mathbf{i}(0, -1, 1) = \mathbf{u} + \mathbf{i}\mathbf{v} .$$

From

$$\mathrm{M}\mathbf{u} = (2, 3, 1) = 2\mathbf{u} - \mathbf{v} , \quad \mathrm{M}\mathbf{v} = (1, -1, 3) = \mathbf{u} + 2\mathbf{v} ,$$

we obtain

$$\mathrm{M}(\alpha\mathbf{u} + \beta\mathbf{v}) = (2\alpha + \beta)\mathbf{u} + (2\beta - \alpha)\mathbf{v}.$$

The subspace spanned by $\mathbf{u}$ and $\mathbf{v}$ is invariant under the action of M.

The algebraic multiplicity *am* of a complex eigenvalue r_k is the exponent of $r - r_k$ in the characteristic polynomial of M .
The geometric multiplicity of r_k is $gm = q - \mathrm{rank}(r_k I - M)$. Notice that now the entries of $r_k I - M$ are complex numbers. Consequently, the determinant is a complex number and it is 0 if and only if its real part and its imaginary part are 0. *gm* coincides with the number of linearly independent pairs of vectors which satisfy (3.3.14).
We have $am \geq gm$, just as in the case of real eigenvalues, and r_k is called semisimple if $am = gm$, and simple if $am = gm = 1$.

The following proposition is very important for future developments.

Proposition 3.3.2 *Eigenvectors of* M *and pairs* $\mathbf{u}, \mathbf{v}$ *corresponding to different eigenvalues are linearly independent.*

Proof. Let us see why this is true for eigenvectors $\mathbf{x}_1, \mathbf{x}_2, ..., \mathbf{x}_q$ corresponding to simple real eigenvalues $r_1, r_2, ..., r_q$.

The result is obviously true for $q = 1$. Assume that it is true for $i < q$ and let $\mathbf{x}_1, \mathbf{x}_2, ..., \mathbf{x}_i$, be linearly independent eigenvectors corresponding to the eigenvalues $r_1, r_2, ..., r_i$. Let $\mathbf{x}_{i+1} \neq \mathbf{0}$ be such that $\mathrm{M}\mathbf{x}_{i+1} = r_{i+1}\mathbf{x}_{i+1}$ and suppose that

$$(3.3.15) \qquad a_1\mathbf{x}_1 + a_2\mathbf{x}_2 + ... + a_i\mathbf{x}_i + a_{i+1}\mathbf{x}_{i+1} = \mathbf{0} .$$

We want to show that all coefficients are 0, i.e. $\mathbf{x}_1, \mathbf{x}_2, ..., \mathbf{x}_i, \mathbf{x}_{i+1}$ are linearly independent. Multiply (3.3.15) by r_{i+1} and obtain

$$(3.3.16) \qquad r_{i+1}a_1\mathbf{x}_1 + r_{i+1}a_2\mathbf{x}_2 + ... + r_{i+1}a_i\mathbf{x}_i + r_{i+1}\, a_{i+1}\mathbf{x}_{i+1} = \mathbf{0} .$$

Applying M to both sides of (3.3.15) gives

$$(3.3.17) \qquad r_1a_1\mathbf{x}_1 + r_2a_2\mathbf{x}_2 + ... + r_ia_i\mathbf{x}_i + r_{i+1}\, a_{i+1}\mathbf{x}_{i+1} = \mathbf{0}\,.$$

Subtract (3.3.17) from (3.3.16). The result is

$$(3.3.18) \qquad (r_{i+1} - r_1)a_1\mathbf{x}_1 + (r_{i+1} - r_2)a_2\mathbf{x}_2 + ... + (r_{i+1} - r_i)a_i\mathbf{x}_i = \mathbf{0}\,.$$

Since $\mathbf{x}_1, \mathbf{x}_2, ..., \mathbf{x}_i$, are linearly independent, all coefficients in (3.3.18) must be 0. This implies $a_1 = ... = a_i = 0$.

From these equalities and (3.3.15) we conclude that a_{i+1} must also be zero. The independence of the eigenvectors is established by induction.

Remark 3.3.2 There is an important and very useful extension of equality (3.3.6). Assume that **X** is a subspace of $\mathbf{R}^q$ such that $M\mathbf{x} \in \mathbf{X}$ for all $\mathbf{x} \in \mathbf{X}$, i.e. **X** is invariant under the action of M. Define

$$\|M\|_X = \sup\{\|M\mathbf{x}\| : \|\mathbf{x}\| = 1, \mathbf{x} \in \mathbf{X}\}\,.$$

Denote by $\rho(M)_X \subset \text{Spec}M$ the subset of eigenvalues such that

- the corresponding eigenvectors are in **X** if the eigenvalue r is real;
- the pairs of vectors **u**, **v** solution of (3.3.14) are in **X** if r is complex.

Then

$$(3.3.19) \qquad \rho(M)_X = \lim_{n\to\infty} \sqrt[n]{\|M^n\|_X}\,.$$

This result will be very useful in the next chapter.

5. **Real canonical form**

In this section we consider matrices M such that all eigenvalues of M are simple, or more generally, semisimple. We first examine the case when they are simple. Later we shall describe how to proceed in the more general case.

Assume that

are the eigenvalues of M, with $r_1, r_2, ..., r_j$ real. We want to show that there is an invertible matrix T such that the matrix $D = T^{-1}MT$ has the following form.

The first row is	r_1 0 0 0 ....0 .
The second row is	0 r_2 0 0 ....0 .
....	
The p-th row is	0 0 0 0 ....0 r_p 0....................................0 .
The (p+1)-th row is	0 0 0 0.........0 a_1 b_1 0 ....0 .
The (p+2)-th row is	0 0 0 0.........0 $-b_1$ a_1 0 ....0 .
....	
The second to the last row is	0 0 0 ....0 $a_{(q-j)/2}$ $b_{(q-j)/2}$.
The last row is	0 0 0 ....0 $-b_{(q-j)/2}$ $a_{(q-j)/2}$.

As entries of the first column of T choose the components of an eigenvector $\mathbf{x}_1$ corresponding to r_1. Similarly, as entries of the second, ..., j-th column of T choose the components of eigenvectors $\mathbf{x}_2, ..., \mathbf{x}_j$ corresponding to $r_2, ..., r_j$.

The entries of the columns $p + 1$ and $p + 2$ of T are the components of a pair of vectors $\mathbf{u}$ and $\mathbf{v}$ found by solving (3.3.12). Continue in this manner until the last two columns of T are in place. The matrix T is invertible, since its column vectors are linearly independent (see Proposition 3.2.1 and Theorem 3.2.1).

It is easily seen that TD and MT are dynamically equal, since for every vector $\mathbf{e}_i$ of the orthonormal basis E we have $T D\mathbf{e}_i = MT\mathbf{e}_i$. In fact, notice that $\mathbf{x}_i = T\mathbf{e}_i$ is the i-th column of the matrix T and assume for simplicity that $\mathbf{x}_i$ is an eigenvector corresponding to a real eigenvalue r_i. Then,

$$MT\mathbf{e}_i = M\mathbf{x}_i = r_i\mathbf{x}_i .$$

From $D\mathbf{e}_i = r_i\mathbf{e}_i$ we obtain

$$TD\mathbf{e}_i = Tr_i\mathbf{e}_i = r_iT\mathbf{e}_i = r_i\mathbf{x}_i = MT\mathbf{e}_i ,$$

which implies that TD and MT are dynamically equal.

Since T is invertible, $D = T^{-1}MT$. The matrix D is called a **real canonical form** of M. Notice that D is not unique. A different real canonical form can be obtained, for example, with a permutation of the real eigenvalues of M along the main diagonal. This permutation will obviously change the form of T.

In the case when the eigenvalues are semisimple, each real eigenvalue is repeated a number of times equal to its algebraic multiplicity. Similarly each block corresponding to a pair of complex conjugate multiple eigenvalues is repeated as many times as the algebraic multiplicity of one of them (their multiplicity is the same). The columns of T are filled by eigenvectors and pairs as in the previous construction, taking care to list first the m_1 linearly independent eigenvectors corresponding to r_1, then the m_2 linearly independent eigenvectors corresponding to r_2, ..., the m_{j+1} pairs corresponding to $r_{j+1} = a_1 + ib_1$, $r_{j+2} = a_1 - ib_1$, etc. . Since for each eigenvalue the algebraic and geometric multiplicity are the same, all columns of T will be filled when the process is completed.

Example 3.3.9 Let

$$M = \begin{pmatrix} 1 & 0 & 1 \\ 0 & 2 & 1 \\ 0 & -1 & 2 \end{pmatrix}.$$

We have seen in the previous example that the eigenvalues of M are $r_1 = 1$, $r_2 = 2 + \mathbf{i}$, $r_3 = 2 - \mathbf{i}$. An eigenvector corresponding to 1 is (1, 0, 0), while for r_2 and r_3 we obtained the two linearly independent vectors $\mathbf{u} = (1, 1, 1)$, $\mathbf{v} = (0, -1, 1)$. Hence

$$D = \begin{pmatrix} 1 & 0 & 0 \\ 0 & 2 & 1 \\ 0 & -1 & 2 \end{pmatrix}, \qquad T = \begin{pmatrix} 1 & 1 & 0 \\ 0 & 1 & -1 \\ 0 & 1 & 1 \end{pmatrix}.$$

It can be easily verified that TD = MT.

Example 3.3.10 Let $M = \begin{pmatrix} 0.5 & 2 \\ 0 & 0 \end{pmatrix}$.

The eigenvalues of M are 0.5 and 0 with corresponding eigenvectors $\mathbf{u} = (1, 0)$ and $\mathbf{v} = (4, -1)$. Consequently

$$T = \begin{pmatrix} 1 & 4 \\ 0 & -1 \end{pmatrix},$$

and a real canonical form of M is

$$D = \begin{pmatrix} 0.5 & 0 \\ 0 & 0 \end{pmatrix}.$$

Example 3.3.11 In Example 3.3.7 we found that the eigenvalues of the matrix

$$M = \begin{pmatrix} 3 & 0 & 1 \\ 0 & 4 & 0 \\ 0 & 0 & 4 \end{pmatrix}$$

are $r_1 = 3$ and $r_2 = r_3 = 4$. An eigenvector corresponding to r_1 is $(1, 0, 0)$ and two linearly independent eigenvectors corresponding to r_2 are $(0, 1, 0)$ and $(1, 0, 1)$. Therefore a real canonical form of M is

$$D = \begin{pmatrix} 3 & 0 & 0 \\ 0 & 4 & 0 \\ 0 & 0 & 4 \end{pmatrix}.$$

Consequently, the matrix T could be

$$T = \begin{pmatrix} 1 & 0 & 1 \\ 0 & 1 & 0 \\ 0 & 0 & 1 \end{pmatrix}.$$

Notice that D is not unique. We could choose

$$D_1 = \begin{pmatrix} 4 & 0 & 0 \\ 0 & 4 & 0 \\ 0 & 0 & 3 \end{pmatrix}.$$

This choice gives, for example,

$$T_1 = \begin{pmatrix} 0 & 1 & 1 \\ 1 & 0 & 0 \\ 0 & 1 & 0 \end{pmatrix}.$$

We could also choose

$$T_2 = \begin{pmatrix} 1 & 0 & 1 \\ 0 & 1 & 0 \\ 1 & 0 & 0 \end{pmatrix}.$$

Section 4. Derivative and First Order Approximation

1. Derivative

Let U be an open subset of $\mathbf{R}^q$, $F : U \to \mathbf{R}^q$ be continuous, and let $\mathbf{y}_0 \in U$. Assume that all partial derivatives of the component functions of F exist in U and are continuous at the point $\mathbf{y}_0$. The matrix

$$M = \begin{pmatrix} \frac{\partial F_1}{\partial x_1} & \frac{\partial F_1}{\partial x_2} & \cdots & \frac{\partial F_1}{\partial x_q} \\ \cdots & \cdots & \cdots & \cdots \\ \frac{\partial F_q}{\partial x_1} & \frac{\partial F_q}{\partial x_2} & \cdots & \frac{\partial F_q}{\partial x_q} \end{pmatrix}$$

where all partial derivatives are evaluated at $\mathbf{y}_0$, is called the **derivative** of F at $\mathbf{y}_0$ and it is denoted by $F_{\mathbf{x}}(\mathbf{y}_0)$. It can be shown that

$$(3.4.1) \qquad \lim_{\|\mathbf{x} - \mathbf{y}_0\| \to 0} \frac{F(\mathbf{x}) - F(\mathbf{y}_0) - M(\mathbf{x} - \mathbf{y}_0)}{\|\mathbf{x} - \mathbf{y}_0\|} = \mathbf{0} ,$$

with $M = F_{\mathbf{x}}(\mathbf{y}_0)$. The equality says that the **norm** of $F(\mathbf{x}) - F(\mathbf{y}_0) - M(\mathbf{x} - \mathbf{y}_0)$ goes to zero faster than the distance $\|\mathbf{x} - \mathbf{y}_0\|$.

Example 3.4.1 Let $\mathbf{x} = (x, y)$, $F(\mathbf{x}) = F(x, y) = (x^2 + y, y^2 - 2x + 1)$. Then, $F_1(x, y) = x^2 + y$, $F_2(x, y) = y^2 - 2x + 1$ and

$$F_{\mathbf{x}}(\mathbf{y}_0) = \begin{pmatrix} 2x_0 & 1 \\ -2 & 2y_0 \end{pmatrix}.$$

Let $\mathbf{y}_0 = (1, 1)$. The derivative of F at $\mathbf{y}_0$ is

$$M = \begin{pmatrix} 2 & 1 \\ -2 & 2 \end{pmatrix},$$

and

$$F(\mathbf{x}) - F(\mathbf{y}_0) - M(\mathbf{x} - \mathbf{y}_0) = ((x - 1)^2, (y - 1)^2).$$

We see that as $\mathbf{x} \to (1, 1)$ the norm of the vector $((x - 1)^2, (y - 1)^2)$ goes to 0 faster than $\|\mathbf{x} - \mathbf{y}_0\| = \sqrt{(x - 1)^2 + (y - 1)^2}$.

2. First order approximation

Guided by the above observation, we call *higher order terms* the difference

$$F(\mathbf{x}) - F(\mathbf{y}_0) - M(\mathbf{x} - \mathbf{y}_0),$$

and we denote it by $o(\|\mathbf{x} - \mathbf{y}_0\|)$. Therefore

$$(3.4.2) \qquad F(\mathbf{x}) = F(\mathbf{y}_0) + M(\mathbf{x} - \mathbf{y}_0) + o(\|\mathbf{x} - \mathbf{y}_0\|).$$

This formula is useful for approximation purposes, since $o(\|\mathbf{x} - \mathbf{y}_0\|)$ becomes negligible and can frequently be omitted when $\mathbf{x}$ is very close to $\mathbf{y}_0$. The omission of the term $o(\|\mathbf{x} - \mathbf{y}_0\|)$ in (3.4.2) destroys the equality and we have

$$(3.4.3) \qquad F(\mathbf{x}) \neq F(\mathbf{y}_0) + M(\mathbf{x} - \mathbf{y}_0),$$

but the right-hand side is a good approximation of $F(\mathbf{x})$ when $\mathbf{x}$ is close to $\mathbf{y}_0$. The right-hand side of (3.4.3) is called **first order approximation** of F at the point $\mathbf{y}_0$. We write

$$(3.4.4) \qquad F(\mathbf{x}) \approx F(\mathbf{y}_0) + M(\mathbf{x} - \mathbf{y}_0), \ M = (F_{\mathbf{x}}(\mathbf{y}_0).$$

Example 3.4.2 With F and $\mathbf{y}_0$ as in Example 3.4.1 we obtain

$$F(\mathbf{x}) \approx (2, 0) + M(x - 1, y - 1) = (2x + y - 1, -2x + 2y).$$

The right-hand side of the equality is the first order approximation of F at $(1, 1)$.

Equality (3.4.1) may hold even in cases when some or all partial derivatives of the component functions are not continuous at $\mathbf{y}_0$. Regardless of the reason behind its validity, we say that F is differentiable at $\mathbf{y}_0$ whenever (3.4.1) holds and we call $F_{\mathbf{x}}(\mathbf{y}_0)$ the derivative of F at $\mathbf{y}_0$. As we move $\mathbf{y}_0$ in U we normally find points where (3.4.1) holds and points where it does not. We say that F is **differentiable** if (3.4.1) holds at every point of U.

As $\mathbf{y}_0$ is moved in U the matrix $F_{\mathbf{x}}(\mathbf{y}_0)$ changes. We say that $F_{\mathbf{x}}(\mathbf{y})$ is continuous at $\mathbf{y}_0$, if for every real number $r > 0$ we can find another real number $s > 0$, such that the operator norm of the matrix $F_{\mathbf{x}}(\mathbf{z}) - F_{\mathbf{x}}(\mathbf{y}_0)$ is smaller than r whenever $\mathbf{z}$ and $\mathbf{y}_0$ are closer than s, i.e.

$$\|\mathbf{z} - \mathbf{y}_0\| < s \text{ implies } \|F_{\mathbf{x}}(\mathbf{z}) - F_{\mathbf{x}}(\mathbf{y}_0)\| < r.$$

It may happen that $F_x(y)$ is continuous at every $y \in U$. In this case we say that F is (of class) C^1. One condition which insures this property is the continuity in U of all partial derivatives of the component functions.

Example 3.4.3 Let F(x, y) = (xCosy, Sin x + Sin y) and let $y_0 = (0, 0)$. The matrix $F_x(y)$ is

$$F_x(y) = \begin{pmatrix} \cos y & -x\sin y \\ \cos x & \cos y \end{pmatrix}.$$

Thus

$$F_x(0) = \begin{pmatrix} 1 & 0 \\ 1 & 1 \end{pmatrix}.$$

The function F is C^1 since all partial derivatives of the component functions are continuous everywhere. The first order approximation of F at (0, 0) is

$$(x\cos y, \sin x + \sin y) \approx (0, 0) + F_x(0)(x, y) = (x, x + y).$$

The first order approximation of a function at some point y_0 of its domain is an important tool for analyzing the behavior of the function close to y_0. As we shall see later certain features of F are completely determined by its first order approximation. The technique of replacing F with its first order approximation in order to detect these features is called **linearization.**

The name *first order approximation* of F given to $F(y_0) + F_x(y_0)(x - y_0)$ is due to the inclusion in it of only the constant vector $F(y_0)$ and the linear terms $F_x(y_0)(x - y_0)$. As $\|x - y_0\| \to 0$, each coordinate of $o(\|x - y_0\|)$ goes to 0 **faster than** the corresponding coordinate of $x - y_0$. For this reason the components of $o(\|x - y_0\|)$ are sometimes called *higher order terms*.

Problems

Section 1.

3.1.1 Let $\mathbf{u} = (1, -1)$, $\mathbf{v} = (2, 1)$. Compute

$2\mathbf{u} + 3\mathbf{v}$, $\mathbf{u} - 2\mathbf{v}$, $\|\mathbf{u}\|$, $\|\mathbf{v}\|$, $\|\mathbf{u} + \mathbf{v}\|$, $\|\mathbf{u} - \mathbf{v}\|$ and $\mathbf{u} \cdot \mathbf{v}$.

3.1.2 Let $\mathbf{u} = (1, 2, -1, 0)$, $\mathbf{v} = (-1, 3, 2, 1)$. Compute

$2\mathbf{u} + 3\mathbf{v}$, $\mathbf{u} - 2\mathbf{v}$, $\|\mathbf{u}\|$, $\|\mathbf{v}\|$, $\|\mathbf{u} + \mathbf{v}\|$, $\|\mathbf{u} - \mathbf{v}\|$ and $\mathbf{u} \cdot \mathbf{v}$.

3.1.3 Verify the Cauchy-Schwarz Inequality for the pairs of vectors $\mathbf{u}$ and $\mathbf{v}$ of Problems 3.1.1 and 3.1.2.

3.1.4 Verify that the two vectors $\mathbf{u} = (1, 2, -1)$, $\mathbf{v} = (-1, 3, 2)$ are linearly independent. Can the vector $(1, 0, 0)$ be written as a linear combination of them?

3.1.5 Verify that the two vectors $\mathbf{u} = (1, -1, 0, 1)$, $\mathbf{v} = (0, 1, -1, 2)$ are linearly independent. Can the vector $(1, 1, 1, 1)$ be written as a linear combination of the vectors $\mathbf{u}$ and $\mathbf{v}$?

3.1.6 Verify that the three vectors $(1, 1)$, $(-1, 2)$, $(2, 3)$ are linearly dependent.

3.1.7 The equation $x + y - z = 0$ represents a subspace of $\mathbf{R}^3$ of dimension 2, in the sense that there is a subspace $\mathbf{V}$ of dimension 2 such that $\mathbf{x} = (x, y, z) \in \mathbf{V}$ if and only if $x + y - z = 0$. Write $\mathbf{V}$ as the span of two linearly independent vectors.

3.1.8 The equation $2x + y - z = 0$ represents a 2-dimensional subspace $\mathbf{W}$ of $\mathbf{R}^3$ (see the previous problem). Write $\mathbf{W}$ as the span of orthogonal vectors.

3.1.9 A subspace of $\mathbf{R}^q$ of dimension 1 is called a line. Let $n = 3$. Verify that the system of equations $x = z, y = 2z$ represents a line of $\mathbf{R}^3$ (in the same sense of the previous two problems).

3.1.10 Let $\mathbf{x}, \mathbf{y}$ be two vectors of $\mathbf{R}^q$. The closed segment from $\mathbf{x}$ to $\mathbf{y}$ is the set of points $\mathbf{z}$ such that $\mathbf{z} = \mathbf{x} + t(\mathbf{y} - \mathbf{x}), t \in [0, 1]$ and it is denoted by $[\mathbf{x}, \mathbf{y}]$.

Let $\mathbf{x} = (1, 2, 3)$, $\mathbf{y} = (-1, 1, 2)$. Verify that the segment $[\mathbf{x}, \mathbf{y}]$ is made with those points $\mathbf{z}$, such that $\mathbf{z} = (1 - 2t, 2 - t, 3 - t)$, $t \in [0, 1]$. Does the vector $(1, 1, 1)$ belong to the segment $[\mathbf{x}, \mathbf{y}]$?

3.1.11 Let $\mathbf{x} = (1, 2)$, $\mathbf{y} = (-1, 1)$. Does the segment $[\mathbf{x}, \mathbf{y}]$ contain a point with equal coordinates?

Section 2.

3.2.1 Let

$$A = \begin{pmatrix} 2 & -3 \\ 1 & 3 \end{pmatrix}, \qquad B = \begin{pmatrix} 1 & 3 \\ -2 & 2 \end{pmatrix}.$$

Find $A + B$, $2A$, $-3B$, AB, BA, A^T and B^T.

3.2.2 Using the property $(AB)^T = B^TA^T$ verify that AA^T is symmetric.

3.2.3 How many diagonal matrices of size 3 can be made with the three numbers 2, 3, 4 if all three must appear as elements of the main diagonal ?

3.2.4 Compute the determinant of the matrix B of Problem 3.2.1. Compute the inverse of B and the inverse of B^2.

3.2.5 Compute the eigenvalues of the matrix A of Problem 3.2.1 and find the spectral radius of A.

3.2.6 Let

$$M = \begin{pmatrix} 1 & -1 & 0 \\ 1 & 0 & -1 \\ 2 & -1 & -1 \end{pmatrix}$$

Verify that $\det M = 0$ and find the rank of M.

3.2.7 Find the eigenvalues and the spectral radius of the matrix M of Problem 3.2.6.

3.2.8 Let

$$M = \begin{pmatrix} 1 & 1 & 1 \\ a & b & c \\ a^2 & b^2 & c^2 \end{pmatrix}.$$

Verify that $\det A = (a - b)(b - c)(c - a)$.

3.2.9 Are the two matrices A and B of Problem 3.2.1 similar ?

Section 3.

3.3.1 Let

$$M = \begin{pmatrix} 1 & -1 & 0 \\ 1 & 0 & -1 \\ 2 & -1 & -1 \end{pmatrix}.$$

Compute $M(1, 0, 2)$, $M(1, 2, -1)$, and in general, $M(x, y, z)$.

3.3.2 Let M be as in the previous problem. Verify that $\mathbf{x} \in \mathrm{Ker}M$ if and only if $x = y = z$. Derive from this result that $\dim\mathrm{Ker}M = 1$. How much is the dimension of ImM ?

3.3.3 Let M be as in Problem 3.3.1. Find the eigenvectors of M. The eigenvalues of M have been already computed in Problem 3.2.7. Find a basis of ImM.

3.3.4 Let

$$C = \begin{pmatrix} 1 & -1 \\ 0 & 1 \end{pmatrix}.$$

Verify that 1 is the only eigenvalue of C. Verify that its algebraic multiplicity is 2, while his geometric multiplicity is 1.

3.3.5 Using the result of Problem 3.2.5, find a real canonical form of the matrix A proposed on that problem.

3.3.6 Let

$$N = \begin{pmatrix} -3 & -2 & 0 \\ -2 & 0 & 0 \\ 0 & 0 & 1 \end{pmatrix}.$$

Find the eigenvalues of N and verify that each one of them has the same algebraic and geometric multiplicity.

3.3.7 Find a real canonical form of the matrix N of Problem 3.3.6.

3.3.8 Let N be the matrix of Problem 3.3.6. Verify that for every $\mathbf{x} \in \mathbf{R}^3$ we have $N^2(\mathbf{x}) = N(N(\mathbf{x}))$.

3.3.9 Let A and B be the matrices of Problem 3.2.1. Consider the action of A in $\mathbf{R}^2$ followed by the action of B, namely the composition $B(A\mathbf{x})$, $\mathbf{x} \in \mathbf{R}^2$. Verify that $B(A\mathbf{x}) = C\mathbf{x}$, where $C = BA$.

3.3.10 Let M, P be two matrices of size q. Justify the inequality $\|MP\| \le \|M\|\,\|P\|$.

3.3.11 Let

$$M = \begin{pmatrix} 1 & 1 \\ 0 & 2 \end{pmatrix}.$$

Find the operator norm of M^n using the Euclidean norm.

3.3.12 Verify that TD = MT in Example 3.3.11. How many different forms of D can be provided ?

Section 4.

3.4.1 Let $F(\mathbf{x}) = (x - y^2, x^2 - y)$. What are the component functions of F? Is the function F continuous in $\mathbf{R}^2$?

3.4.2 Let F be the function of Problem 3.4.1 . Find the derivative of F at $\mathbf{x} = (1, 2)$

and use it to compute the first order approximation of F at the point (1, 2). Verify the limit (3.4.1).

3.4.3 Let $F(\mathbf{x}) = (x - y + z, x + z, y + z)$. Verify that F can be represented by a matrix, in the sense that there exists a matrix M such that $M(\mathbf{x}) = F(\mathbf{x})$ for all $\mathbf{x} \in \mathbf{R}^3$. Verify that the derivative of F is exactly the matrix M.

3.4.4 Let $F(\mathbf{x}) = (\frac{x}{y}, \frac{y}{x})$. Is the function F defined everywhere in $\mathbf{R}^2$? Find the derivative of F in general, and at the point (2, 3).

3.4.5 Derive the first order approximation of the function F of Problem 3.4.4 at the point (2, 3) .

CHAPTER 4

DISCRETE LINEAR DYNAMICAL SYSTEMS

SUMMARY

The purpose of this chapter is to study linear dynamical systems in a dimension higher than 1. The first section contains some general information about linear systems. In the second section we study the stability of the origin, which is always an equilibrium point of any linear system. Using the eigenvalues of a matrix, we determine under which conditions all orbits converge to the origin (all eigenvalues have modulus smaller than 1), or all non stationary orbits diverge to infinity (all eigenvalues have modulus larger than 1). In the third section we examine the case when some eigenvalues have modulus smaller than 1 and the others have modulus larger than 1. In this section we introduce the idea of stable and unstable subspace. Finally, in section 4, we analyze the case when there are eigenvalues with modulus equal to 1.

Section 1. Linear and Affine Processes

A linear operator $L : \mathbf{R}^q \to \mathbf{R}^q$ is a function from $\mathbf{R}^q$ into itself such that

$$(4.1.1) \qquad L(a\mathbf{x} + b\mathbf{y}) = aL\mathbf{x} + bL\mathbf{y}$$

for every pair of vectors $\mathbf{x}, \mathbf{y} \in \mathbf{R}^q$ and every pair a, b of real numbers.
The matrix $M = \{m_{ij}, q\}$ such that the entry m_{ij} is the j-th component of the vector $L\mathbf{e}_i$, for all $1 \le i, j \le q$, has the important property

$$(4.1.2) \qquad L\mathbf{x} = M\mathbf{x},$$

for every $\mathbf{x} \in \mathbf{R}^q$, when we use the standard basis E. In fact, from $\mathbf{x} = x_1\mathbf{e}_1 + \ldots + x_q\mathbf{e}_q$ and using (4.1.1) we obtain

$$L\mathbf{x} = x_1L\mathbf{e}_1 + \ldots + x_qL\mathbf{e}_q.$$

The i-th component of $L\mathbf{x}$ is exactly $\mathbf{m}_i \cdot \mathbf{x}$ where $\mathbf{m}_i$ is the i-th row of M.

A **linear** ($\mathbf{b} = \mathbf{0}$) or **affine** dynamical system in $\mathbf{R}^q$ has the form

$$(4.1.3) \qquad \mathbf{x}_{n+1} = L\mathbf{x}_n + \mathbf{b}$$

where $L : \mathbf{R}^q \to \mathbf{R}^q$ is a linear operator. Using (4.1.2) and the standard basis of $\mathbf{R}^q$, we can write (4.1.3) in the form

$$(4.1.4) \qquad \mathbf{x}_{n+1} = M\mathbf{x}_n + \mathbf{b}$$

where M is the matrix defined above. Thus, the study of discrete linear dynamical systems in $\mathbf{R}^q$ can be reduced to the analysis of (4.1.4). This is our approach in the present chapter.

Example 4.1.1 Let $L\mathbf{x} = L(x, y) = (3x + 2y, -x + y)$.
Then $L\mathbf{e}_1 = (3, -1)$ and $L\mathbf{e}_2 = (2, 1)$. The matrix M is

$$M = \begin{pmatrix} 3 & 2 \\ -1 & 1 \end{pmatrix}.$$

It is easily seen that $Lx = Mx$ for every $x \in R^2$, $x = xe_1 + ye_2$.

Let $Nx = Mx + b$. Then, given $x_0 \in R^q$, we have

$$x_1 = Nx_0 = Mx_0 + b, \ \ x_2 = Nx_1 = Mx_1 + b = M^2x_0 + Mb + b$$

and in general,

$$x_{n+1} = Nx_n = M^{n+1}x_0 + (M^n + M^{n-1} + \dots + M + I)b \, . \tag{4.1.5}$$

Example 4.1.2 Let $b = (1, 1)$ and

$$M = \begin{pmatrix} 1 & 2 \\ 1 & 1 \end{pmatrix}.$$

We have

$$x_1 = N(x_0) = N(x_0, y_0) = M(x_0, y_0) + (1, 1) =$$

$$(x_0 + 2y_0, x_0 + y_0) + (1, 1) = (x_0 + 2y_0 + 1, x_0 + y_0 + 1);$$

$$x_2 = N(x_1) = (x_1 + 2y_1 + 1, x_1 + y_1 + 1) = M(x_1, y_1) + (1, 1) =$$

$$M^2(x_0, y_0) + M(1, 1) + (1, 1) \text{ etc.}$$

Using the idea of conjugate maps we can reduce the study of (4.1.4) (and (4.1.5)) to the analysis of the simpler system governed only by the matrix M in the case when 1 is not an eigenvalue of M.

In other words, we construct an invertible map $B : R^q \to R^q$ such that

$$N(B(x)) = B(Mx) \, , \tag{4.1.6}$$

for all $x \in R^q$, provided that $1 \notin \text{Spec}M$.

We are interested in obtaining equality (4.1.6) in the simplest possible manner. Therefore, we try $Bx = Ax + c$ where A is a matrix. Substitution in (4.1.6) gives

$$N(B(x)) = M(Ax+c) + b = MAx + Mc + b = B(Mx) = AMx + c \, . \tag{4.1.7}$$

Equality (4.1.7) must hold for every x. Hence

$$AM = MA, \text{ and } Mc + b = c \, .$$

The simplest way to obtain the first equality is to choose $A = I$, the identity matrix. Moreover, since 1 is not an eigenvalue of M, there exists one and only one vector $\mathbf{c}$ such that

$$(4.1.8) \qquad M\mathbf{c} + \mathbf{b} = \mathbf{c}.$$

In fact, from (4.1.8) we get $\mathbf{c} = (I - M)^{-1}\mathbf{b}$.

The effects of the presence of 1 among the eigenvalues of the matrix M are studied in the fourth section of this chapter. Hence, for the time being, the study of the affine dynamical system (4.1.4) is reduced to the study of the linear system

$$(4.1.9) \qquad \mathbf{x}_{n+1} = M\mathbf{x}_n \ ,$$

by means of the conjugacy $B(\mathbf{x}) = \mathbf{x} + \mathbf{c}$, where $\mathbf{c}$ is the unique solution of (4.1.8).

Notice that $M\mathbf{0} = \mathbf{0}$. Hence the origin is always an equilibrium point of (4.1.9). No other equilibrium point exists in our case since $1 \notin \mathrm{Spec}M$, which implies that the equation

$$(4.1.10) \qquad M\mathbf{x} = \mathbf{x}$$

has only the trivial solution $\mathbf{x} = \mathbf{0}$. From equality (4.1.6) we derive that $\mathbf{c} = B(\mathbf{0})$ is the only stationary state of the dynamical system governed by N.

In the next two sections, we shall study the properties of the origin as the unique equilibrium point of the dynamical system (4.1.9). Consequently, we obtain the properties of $\mathbf{c} = B(\mathbf{0})$ as the unique stationary state of (4.1.4). The reader may ask what about **periodic orbits** ? As we shall see the existence of periodic orbits requires the presence of complex conjugate eigenvalues with modulus 1 in the spectrum of M. In Section 4 we will consider this case as well.

Example 4.1.3 Let $\mathbf{b} = (-3, 2)$ and

$$M = \begin{pmatrix} 2 & 1 \\ 0 & -1 \end{pmatrix}.$$

The eigenvalues of M are 2 and -1. To find the conjugacy B between the dynamical system generated by $\mathbf{x}_{n+1} = M\mathbf{x}_n + \mathbf{b}$ and the system $\mathbf{y}_{n+1} = M\mathbf{y}_n$ we need to find the vector $\mathbf{c}$.

Let **c** = (x, y). From (4.1.8) we derive the equality

$$(x, y) - M(x, y) = (-3, 2),$$

which gives the system of two equations in two unknowns

$$\begin{cases} x + y = 3 \\ 2y = 2. \end{cases}$$

Hence y = 1, x = 2 and **c** = (2, 1).

The only stationary state of the dynamical system governed by M is the origin. Therefore the only stationary state of the system governed by N is the point

$$B(0, 0) = (0, 0) + (2, 1) = (2, 1) = \mathbf{c}.$$

Example 4.1.4 Let **b** = (a, b) and let

$$M = \begin{pmatrix} 2 & 1 \\ 0 & 3 \end{pmatrix}.$$

From (x, y) - M(x, y) = (a, b) we obtain the system of two equations in two unknowns

$$\begin{cases} x + y = -a \\ 2y = -b. \end{cases}$$

For every choice of (a, b) the system has the only solution $x = -a + b/2$, $y = -b/2$. Hence **c** = (- a + b/2, - b/2) .

Section 2. Stability and Instability of the Origin

1. The origin as a sink

We want to find a condition which insures that every orbit of the dynamical system (4.1.9) converges to **0**. The condition requires knowledge of the spectral radius of M. Numerical methods can be of great help in providing an answer to this technical aspect of the problem, and the interested reader is encouraged to use them for obtaining the necessary information. One of these methods is described in Section 4 of Chapter 6.

The norm we have considered so far in $\mathbf{R}^q$ is the so-called Euclidean norm, but there are cases in which this norm may not be the best, at least for establishing certain results. A case in point is the following Lemma.

Lemma 4.2.1 *Let* M *be a matrix of size* q *such that* $\rho(M) < 1$. *Then we can define in* $\mathbf{R}^q$ *a new norm* $\|\cdot\|_N$ *so that the operator norm of* M *induced by* $\|\cdot\|_N$ *is smaller than* 1.

Proof. First we observe that, according to the equality (see (3.3.6))

$$\rho(M) = \lim_{n\to\infty} \sqrt[n]{\|M^n\|},$$

there exists a positive integer p such that the operator norm of M^p, induced by the Euclidean norm in $\mathbf{R}^q$, is smaller than 1, i.e. $\|M^p\| < 1$. Guided by this observation, we define the new norm in $\mathbf{R}^q$ as follows (the subscript N stands for **new**):

$$\|\mathbf{x}\|_N = \|\mathbf{x}\| + \|M\mathbf{x}\| + \ldots + \|M^{p-1}\mathbf{x}\| . \tag{4.2.1}$$

To check the operator norm of M, when in $\mathbf{R}^q$ one uses this new norm, we pick any vector $\mathbf{x}$ such that $\|\mathbf{x}\|_N = 1$. Then

$$\|M\mathbf{x}\|_N = \|M\mathbf{x}\| + \ldots + \|M^p\mathbf{x}\| = \|\mathbf{x}\| + \|M\mathbf{x}\| + \ldots + \|M^p\mathbf{x}\| - \|\mathbf{x}\| =$$

$$= 1 - \|\mathbf{x}\| + \|M^p\mathbf{x}\| \leq 1 - \|\mathbf{x}\|(1 - \|M^p\|) .$$

From $\|\mathbf{x}\|_N = 1$ and using (4.2.1) we obtain

$$1 \leq (1 + \|M\| + \ldots + \|M^{p-1}\|)\, \|\mathbf{x}\| = K\|\mathbf{x}\| , \tag{4.2.2}$$

with $K = 1 + \|M\| + \ldots + \|M^{p-1}\|$. Thus

$$\|Mx\|_N \le 1 - \frac{1}{K}(1 - \|M^p\|) \quad \text{for every } \mathbf{x} \text{ such that } \|x\|_N = 1.$$

The operator norm $\|M\|_N$, associated to the new norm of $\mathbf{R}^q$, satisfies the inequality

$$(4.2.3) \qquad \|M\|_N \le 1 - \frac{1}{K}(1 - \|M^p\|) < 1 .$$

Remark 4.2.1 Two important observations are in order. First, notice that from (4.2.1), (4.2.2) we derive

$$(4.2.4) \qquad \|x\| \le \|x\|_N \le K\|x\| .$$

In other words the new norm of $\mathbf{x}$ is bounded below by the Euclidean norm of $\mathbf{x}$ and bounded above by the same Euclidean norm multiplied by K.
Second, observe that Lemma 4.2.1 can be extended, with the same proof, to an invariant subspace $\mathbf{X}$ of $\mathbf{R}^q$ provided that $\|M\|$, $\rho(M)$ are respectively replaced by $\|M\|_X$, $\rho(M)_X$ (see Chapter 3, Remark 3.3.2).

Lemma 4.2.1 is all we need to prove the following important result.

Theorem 4.2.1 *Assume that all eigenvalues of* M *have modulus smaller than* 1. *Then every orbit of the dynamical system governed by* M *converges to* $\mathbf{0}$.

Proof. Since $\rho(M)$ is the modulus of the largest eigenvalue of M, we have $\rho(M) < 1$. According to the previous lemma, we can define in $\mathbf{R}^q$ a new norm $\| \cdot \|_N$ in such a way that $\|M\|_N < 1$. From

$$\mathbf{x}_n = M^n \mathbf{x}_0 \quad \text{(see (4.1.5) with } \mathbf{b} = \mathbf{0}),$$

we derive (observe that $\|M^p\|_N \le \|M\|_N^p$)

$$\|M^n \mathbf{x}_0\|_N \le \|M^n\|_N \|\mathbf{x}_0\|_N \le (\|M\|_N)^n \|\mathbf{x}_0\|_N .$$

Since $\|M\|_N < 1$ every trajectory converges to $\mathbf{0}$. One may observe that the convergence is in the new norm, but inequality (4.2.4) implies $\|M^n\mathbf{x}_0\| \le \|M^n\mathbf{x}_0\|_N$. Thus every trajectory converges to $\mathbf{0}$ in the Euclidean norm as well (or in any other norm (see (1.1.15))) .

Example 4.2.1 Let

$$M = \begin{pmatrix} 0.5 & 2 \\ 0 & 0 \end{pmatrix}.$$

We want to study the dynamical system $\mathbf{x}_{n+1} = M\mathbf{x}_n + (0, 2)$.

According to the ideas presented in the previous section we can confine our analysis to the behavior of the dynamical system $\mathbf{x}_{n+1} = M\mathbf{x}_n$. The eigenvalues of M are 0.5 and 0 . Thus every orbit converges to **0**.

We can reach the same conclusion using a canonical form of M. The eigenvectors of M are (1, 0), corresponding to 0.5, and (4, -1). A real canonical form of M is (see Chapter 3)

$$D = \begin{pmatrix} 0.5 & 0 \\ 0 & 0 \end{pmatrix}.$$

The action of the dynamical system governed by D is easily seen:

$$D^n(x, y) = ((\frac{1}{2})^n x, 0).$$

From $M = TDT^{-1}$, and

$$T = \begin{pmatrix} 1 & 4 \\ 0 & -1 \end{pmatrix}, \quad T^{-1} = \begin{pmatrix} 1 & 4 \\ 0 & -1 \end{pmatrix},$$

we obtain

$$M^n(x, y) = TD^nT^{-1}(x, y) = TD^n(x + 4y, -y) = T(2^{-n}(x + 4y), 0) = (2^{-n}(x + 4y), 0).$$

Hence $M^n(x, y) = (2^{-n}(x + 4y), 0) \to (0, 0)$ as $n \to \infty$.

In this case the result could have been established also in a more direct manner since

$$M^n = \begin{pmatrix} (\frac{1}{2})^n & (\frac{1}{2})^{n-2} \\ 0 & 0 \end{pmatrix}, \qquad n = 1, 2, \dots .$$

Going back to our original goal, which was the study of the dynamical system $\mathbf{x}_{n+1} = M\mathbf{x}_n + (0, 2)$ we find $\mathbf{c} = (8, 2)$. Hence $B(\mathbf{x}) = \mathbf{x} + (8, 2)$. The only stationary state of the system is $B(0, 0) = (8, 2)$ and every orbit converges to it. The orbits which collapse immediately into the stationary state are the ones starting at points (x, y) such that $N(x, y) = (8, 2)$, or $(0.5x + 2y, 0) + (0, 2) = (8, 2)$. This implies $x + 4y = 16$. This line goes through the point $(8, 2)$ and it is parallel to the line representing KerM, since $\text{KerM} = \{(x, y) : x + 4y = 0\}$.

Example 4.2.2 Let

$$M = \begin{pmatrix} 1 & -1 \\ 0.5 & -1 \end{pmatrix}.$$

The eigenvalues of M are $\sqrt{0.5}$ and $-\sqrt{0.5}$. Hence every orbit of the dynamical system $\mathbf{x}_{n+1} = M\mathbf{x}_n$ converges to $\mathbf{0}$. Notice that the operator norm of M, using the Euclidean norm in $\mathbf{R}^2$, is not smaller than $\sqrt{2} = \|M(0, 1)\|$. This fact may wrongly be interpreted as an indication that some trajectories of the dynamical system defined by M are not going to $\mathbf{0}$. The reader is encouraged (see Problem 4.2.1 and the previous exercise) to determine a real canonical form of M and to find the form of $M^n(x, y)$, for every $(x, y) \in \mathbf{R}^2$.

2. **The origin as a source**

We now assume that all eigenvalues of M have modulus larger than 1. In this case the matrix is invertible and the inverse satisfies the inequality $\rho(M^{-1}) < 1$.

Therefore, according to Lemma 4.2.1, we can find in $\mathbf{R}^q$ a new norm such that $\|M^{-1}\|_N < 1$, and we obtain

$$\|\mathbf{x}\|_N = \|M^{-1}M\mathbf{x}\|_N \leq \|M^{-1}\|_N \, \|M\mathbf{x}\|_N . \tag{4.2.5}$$

Inequality (4.2.5) tells us that there is a constant $r > 1$ such that $\|M\mathbf{x}\|_N \geq r\,\|\mathbf{x}\|_N$ for every $\mathbf{x} \in \mathbf{R}^q$.

On the basis of this observation we obtain the following important result.

Theorem 4.2.2 *Assume that all eigenvalues of* M *have modulus larger than* 1. *Then for every* $\mathbf{x} \neq \mathbf{0}$ *we have*

$$\|M^n\mathbf{x}\| \to \infty \textit{ as } n \to \infty. \tag{4.2.6}$$

<u>Proof</u>. Let $\mathbf{x}_0 \in \mathbf{R}^q$, $\mathbf{x}_0 \neq \mathbf{0}$. From the inequality

$$\|M\mathbf{x}_0\|_N \geq r\,\|\mathbf{x}_0\|_N, \text{ with } r = \frac{1}{\|M^{-1}\|_N} > 1,$$

we derive $\|M^n x_0\|_N \geq r^n \|x_0\|_N$, which implies that $\|M^n x\|_N \to \infty$ as $n \to \infty$. From the inequality

$$\|x\| \leq \|x\|_N \leq K\|x\| \qquad \text{(see (4.2.4))},$$

we obtain

$$\|M^n x\| \geq K^{-1} r^n \|x_0\| ,$$

which implies (4.2.6).
Hence, every non stationary orbit of the dynamical system $x_{n+1} = Mx_n$ goes to infinity.

Example 4.2.3 Let

$$M = \begin{pmatrix} 2 & 2 \\ 0 & 3 \end{pmatrix}.$$

The eigenvalues of M are 2 and 3. Thus every non stationary trajectory of the dynamical system defined by M goes to infinity.

The two previous results are somehow extreme cases. In the first, every eigenvalue of M has modulus smaller than 1 and every orbit goes to the origin. In the second every eigenvalue of M has modulus larger than 1 and every orbit with initial state $x \neq 0$ goes to infinity.

Section 3. The Origin as a Saddle Point

We now examine the case when some of the eigenvalues of M have modulus larger than 1 and the others have modulus smaller than 1. The stationary state **0** is then called a **saddle point**. To understand the behavior of a dynamical system governed by a matrix having this property we start with a simple example.

Example 4.3.1 Let

$$M = \begin{pmatrix} 2^{-1} & 2 \\ 0 & 3 \end{pmatrix}.$$

The eigenvalues of M are 2^{-1} and 3. Eigenvectors of M are (1, 0), which corresponds to 2^{-1}, and (4, 5), corresponding to 3. The trajectory starting at any vector of the form (t, 0) will go to zero, since $M^n(t, 0) = (2^{-n} t, 0)$. For this reason the span of (1, 0) (in the plane this span is the x-axis) is called the **stable subspace** of M.
Similarly, the trajectory starting at any vector of the form (4 t, 5 t) with $t \neq 0$, goes to infinity, since $M^n(4\,t, 5\,t) = (3^n 4\,t, 3^n 5\,t)$. The span of the vector (4, 5), which is a one-dimensional subspace of $\mathbf{R}^2$, is called the **unstable subspace** of M.

Since $\mathbf{u} = (4, 5)$ and $\mathbf{v} = (1, 0)$ are a basis of $\mathbf{R}^2$, every vector $\mathbf{x} \in \mathbf{R}^2$ can be written as a linear combination of **u** and **v**, $\mathbf{x} = a\mathbf{u} + b\mathbf{v}$, for some real numbers a and b. This implies

$$M^n\mathbf{x} = 3^n a\mathbf{u} + 2^{-n} b\mathbf{v}.$$

As $n \to \infty$ the part $2^{-n}b\mathbf{v}$ goes to zero, while $3^n a\mathbf{u}$ goes to infinity. The orbit starting at any point $\mathbf{x} = a\mathbf{u} + b\mathbf{v}$ with $a \neq 0$ goes to infinity while approaching the unstable subspace of M.

In summary, every vector $\mathbf{x} \in \mathbf{R}^2$ can be written as a sum of two vectors, one (**v**) belonging to the stable subspace of M and the other (**u**) belonging to the unstable subspace of M. Each state of the orbit of **x** is the sum of the corresponding states of the orbits of **u** and **v**. The orbit of **v** goes to **0**, and the orbit of **u** goes to infinity. Each orbit remains in the corresponding subspace. Consequently, the orbit of **x** goes to infinity while approaching the unstable subspace.

Guided by the above example, we will now analyze more complex cases. The results will be presented in an informal manner, rather than stated as theorems.

1. The action of a real canonical form of M

We assume that all eigenvalues of M are simple. Later our reasoning will be adjusted to the case when each eigenvalue of M is semisimple. We also analyze the case when there are eigenvalues which are not semisimple.

Let D be a real canonical form of M, $D = T^{-1}MT$. We first list along the main diagonal of D the real eigenvalues $r_1, r_2, ..., r_j$, of M, starting from the largest and going down in decreasing order. Then we include the entries corresponding to the complex conjugate eigenvalues, starting from the ones with largest modulus and going down to the ones with smallest modulus.

The action of D is completely determined by its behavior on the elements

$$\mathbf{e}_1 = (1, 0, ..., 0),\ \mathbf{e}_2 = (0, 1, ..., 0),\ ...,\ \mathbf{e}_q = (0, 0, ..., 1).$$

Since $D^n\mathbf{e}_1 = (r_1)^n\mathbf{e}_1$, the trajectory starting from each vector of the form $t\mathbf{e}_1$, with $t \neq 0$, goes to infinity if $|r_1| > 1$ and to $\mathbf{0}$ if $|r_1| < 1$. Similar comment holds true for the vectors $\mathbf{e}_2, ..., \mathbf{e}_j$ corresponding to the eigenvalues $r_2, ..., r_j$.

Assume now that $r_{j+1} = a + \mathbf{i}b$ and $r_{j+2} = a - \mathbf{i}b$. Then row j+1 and j+2 of D have the form

$$\begin{matrix} 0 & 0 \ldots\ldots 0 & a & b & 0 \ldots\ldots 0 \\ 0 & 0 \ldots\ldots 0 & -b & a & 0 \ldots\ldots 0 \end{matrix}$$

with $d_{j+1,j+1} = d_{j+2,j+2} = a$, $d_{j+1,j+2} = -d_{j+2,j+1} = b$. Without loss of generality we may assume that, as far as these two eigenvalues are concerned, everything is happening in $\mathbf{R}^2$ and we are dealing with the action of the matrix

$$P = \begin{pmatrix} a & b \\ -b & a \end{pmatrix}.$$

Writing $r = \sqrt{a^2 + b^2}$ we obtain

$$P = r\begin{pmatrix} \frac{a}{r} & \frac{b}{r} \\ \frac{-b}{r} & \frac{a}{r} \end{pmatrix}.$$

Since

$$(\frac{a}{r})^2 + (\frac{b}{r})^2 = 1 ,$$

there exists an angle θ such that

$$\frac{a}{r} = \text{Cos}\,\theta , \quad \text{and} \quad \frac{b}{r} = \text{Sin}\,\theta .$$

Therefore

$$P = r \begin{pmatrix} \text{Cos}\,\theta & \text{Sin}\,\theta \\ -\text{Sin}\,\theta & \text{Cos}\,\theta \end{pmatrix},$$

$$P^n = r^n \begin{pmatrix} \text{Cos}\,n\theta & \text{Sin}\,n\theta \\ -\text{Sin}\,n\theta & \text{Cos}\,n\theta \end{pmatrix}$$

The action of P is a rotation of an angle θ followed by an expansion ($r > 1$) or compression ($r < 1$) of magnitude r.
Therefore, the orbit of every non zero vector of the two-dimensional subspace spanned by the vectors $\mathbf{e}_j$ and $\mathbf{e}_{j+1}$ goes to infinity if $r > 1$ and to $\mathbf{0}$ if $r < 1$. The rotational motion gives the trajectory the appearance of a spiral.

The remaining complex conjugate eigenvalues are analyzed in the same manner.

In summary, under the action of D, the set of vectors of $E = \{\mathbf{e}_1, ..., \mathbf{e}_q\}$ corresponding to real eigenvalues in absolute value smaller than 1, together with the pairs corresponding to complex conjugate eigenvalues with modulus smaller than 1, span a subspace $\mathbf{X}_S$ of $\mathbf{R}^q$, such that a trajectory starting at any point of this subspace goes to $\mathbf{0}$. For this reason $\mathbf{X}_S$ is called the **stable subspace** of the dynamical system defined by D.

Likewise, the vectors of E corresponding to real eigenvalues in absolute value larger than 1, together with the pairs corresponding to complex conjugate eigenvalues with modulus larger than 1, span a subspace $\mathbf{X}_U$ of $\mathbf{R}^q$ such that a trajectory starting at any point of $\mathbf{X}_U$ will go to infinity. For this reason $\mathbf{X}_U$ is called the **unstable subspace** of the dynamical system defined by D.

Example 4.3.2 Let

$$D = \begin{pmatrix} 0.5 & 0 & 0 \\ 0 & 2 & 1 \\ 0 & -1 & 2 \end{pmatrix}.$$

The one-dimensional subspace of $\mathbf{R}^3$ spanned by $\mathbf{e}_1$ is the stable subspace of D. The unstable subspace is the plane spanned by the vectors $\mathbf{e}_2$ and $\mathbf{e}_3$. The action of D on this subspace is a rotation of an angle $\theta = \text{ArcTan}0.5 \cong 26.5^o$ combined with an expansion of $r = \sqrt{5}$. For example, the vector $\mathbf{x} = (0, 1, 1)$ is moved to $D\mathbf{x} = (0, 3, 1)$. We see that

$$\|D\mathbf{x}\|^2 = 10 = 5\,\|\mathbf{x}\|^2, \text{ and } \operatorname{Cos}\theta = \frac{2}{\sqrt{5}}, \text{ i.e. } \theta \cong 26.5^o.$$

All steps and conclusions of the above analysis can be easily adjusted to the case when some real diagonal elements or 2×2 blocks of the matrix D are repeated according to the multiplicity of the corresponding semisimple eigenvalue.

Example 4.3.3 Let

$$D = \begin{pmatrix} .5 & 0 & 0 \\ 0 & .5 & 0 \\ 0 & 0 & 2 \end{pmatrix}.$$

From $D\mathbf{e}_1 = .5\mathbf{e}_1$, $D\mathbf{e}_2 = .5\mathbf{e}_2$, we derive $D\mathbf{x} = .5\mathbf{x}$ for every $\mathbf{x} \in \text{Span}\{\mathbf{e}_1, \mathbf{e}_2\}$. Thus, the stable subspace of D is the plane $z = 0$. The unstable subspace is the z-axis $(x = y = 0)$.

2. The action of M

Assume that all eigenvalues of M are semisimple. Then, as a consequence of the previous analysis on the action of D, we obtain that the behavior of M is completely determined by its action on the vectors $T\mathbf{e}_1, \ldots, T\mathbf{e}_q$ since $TD^n\mathbf{e}_i = M^nT\mathbf{e}_i$. Let us divide these vectors into two groups : G_S and G_U. The first group contains all vectors $T\mathbf{e}_i$ which correspond to real (or complex conjugate) eigenvalues with absolute value (or modulus) smaller than 1. The second group, G_U, contains the vectors $T\mathbf{e}_i$ corresponding to the remaining eigenvalues.

Under the action of M the orbit of a vector of $\mathbf{R}^q$ which can be written as a linear combination of the elements of G_S will converge to $\mathbf{0}$, while the orbit of a non-zero vector which is a linear combination of the elements of G_U goes to infinity.

The subspace $\mathbf{W}_S \subset \mathbf{R}^q$ spanned by the elements of G_S is called the **stable subspace** of M, and the subspace $\mathbf{W}_U \subset \mathbf{R}^q$ spanned by the elements of G_U is called the **unstable subspace** of M. Notice that

$$\mathbf{W}_S = T\mathbf{X}_S, \quad \mathbf{W}_U = T\mathbf{X}_U .$$

Every vector $\mathbf{x}$ of $\mathbf{R}^q$ is a linear combination of the vectors spanning $\mathbf{W}_S$ and $\mathbf{W}_U$. In other words, $\mathbf{x}$ can be written in a unique manner as

$$\mathbf{x} = \mathbf{x}_S + \mathbf{x}_U \ , \tag{4.3.1}$$

with $\mathbf{x}_S \in \mathbf{W}_S$, and $\mathbf{x}_U \in \mathbf{W}_U$.

The orbit of $\mathbf{x}$ is the sum of the two orbits $O(\mathbf{x}_S)$ and $O(\mathbf{x}_U)$, i.e.

$$M^n\mathbf{x} = M^n\mathbf{x}_S + M^n\mathbf{x}_U \ , \tag{4.3.2}$$

Hence, $O(\mathbf{x})$ will go to infinity if and only if the $\mathbf{x}_U$ component of $\mathbf{x}$ is not $\mathbf{0}$.

Example 4.3.4 Let

$$M = \begin{pmatrix} 1 & 2 & 0 \\ -2 & 1 & 1 \\ 0 & 0 & 0.5 \end{pmatrix}.$$

The eigenvalues of M are $r_1 = 1/2$, and $r_2 = 1 + 2\mathbf{i}$, $r_3 = 1 - 2\mathbf{i}$. Since $|r_2| = |r_3| = \sqrt{5}$ we obtain that

$$X_S = \text{Span}\{\mathbf{e}_1\} \text{ and } X_U = \text{Span}\{\mathbf{e}_2, \mathbf{e}_3\}$$

are the stable and unstable subspaces of the real canonical form, D, of M

$$D = \begin{pmatrix} 0.5 & 0 & 0 \\ 0 & 1 & 2 \\ 0 & -2 & 1 \end{pmatrix}.$$

The action of D on the vectors of the form $(t, 0, 0)$ is $D^n(t, 0, 0) = (2^{-n}t, 0, 0)$ while the action of D on vectors of the form $(0, u, v)$ is

$$D^n(0, u, v) = \sqrt{5^n}\ (0, u\text{Cos}\, n\theta + v\ \text{Sin}\, n\theta, - u\ \text{Sin}\, n\theta + v\ \text{Cos}\, n\theta\ ),$$

where $\theta = \mathrm{Cos}^{-1}\frac{1}{\sqrt{5}}$. Hence, after n steps, the vector has been rotated (clockwise) by an angle $n\theta$ and its norm has been multiplied by $\sqrt{5^n}$.

The matrix T such that $D = T^{-1}MT$ is obtained by solving the equations

$$M\mathbf{x} = 0.5\mathbf{x}, \text{ and } M\mathbf{w} = (1 + 2\mathbf{i})\mathbf{w}.$$

A non zero solution of $M\mathbf{x} = \frac{1}{2}\mathbf{x}$ is the vector $\mathbf{x} = (8, -2, 17)$. The second equation

$$M\mathbf{w} = (1+ 2\mathbf{i})\mathbf{w} \text{ is solved by } \mathbf{w} = (1, \mathbf{i}, 0) = (1, 0, 0) + \mathbf{i}(0, 1, 0) = \mathbf{e}_1 + \mathbf{i}\mathbf{e}_2.$$

Consequently,

$$T = \begin{pmatrix} 8 & 1 & 0 \\ -2 & 0 & 1 \\ 17 & 0 & 0 \end{pmatrix}, \text{ and } T^{-1} = \frac{1}{17}\begin{pmatrix} 0 & 0 & 1 \\ 17 & 0 & -8 \\ 0 & 17 & 2 \end{pmatrix}.$$

The stable subspace of M is

$$W_S = \mathrm{Span}\{T\mathbf{e}_1\} = \mathrm{Span}\ \{(8, -2, 17)\}\ .$$

The unstable subspace of M is

$$W_U = \mathrm{Span}\{T\mathbf{e}_2, T\mathbf{e}_3\} = \mathrm{Span}\ \{\mathbf{e}_1, \mathbf{e}_2\}.$$

Let us find the form of the orbit $O(\mathbf{x}) = O((x, y, z))$. Notice that

$$D^n(u, v, w) = \left([\tfrac{1}{2}]^n u\ , \sqrt{5^n}\,(v\mathrm{Cos}\ n\theta + w\mathrm{Sin}\ n\theta\ )\ , \sqrt{5^n}\ (-\ v\ \mathrm{Sin}\ n\theta + w\ \mathrm{Cos}\ n\theta\ )\ \right).$$

Since

$$T^{-1}(x, y, z) = (\tfrac{z}{17}, x - \tfrac{8z}{17}, y + \tfrac{2z}{17})\ .$$

we obtain

$$D^nT^{-1}(x, y, z) = \left([\tfrac{1}{2}]^n \tfrac{z}{17}, \sqrt{5^n}\,[x - \tfrac{8z}{17}]\mathrm{Cos}\ n\theta + \sqrt{5^n}\,[\ y + \tfrac{2z}{17}]\mathrm{Sin}\ n\theta\ ,\right.$$

$$\left. -\ \sqrt{5^n}\,[x - \tfrac{8z}{17}]\mathrm{Sin}\ n\theta + \sqrt{5^n}\,[y + \tfrac{2z}{17}]\mathrm{Cos}\ n\theta \right)\ .$$

From

$$M^n(x,y,z) = TD^nT^{-1}(x, y, z) ,$$

we derive

$$M^n(x,y,z) = T(D^nT^{-1}(x, y, z))= [\tfrac{1}{2}]^n \left(\frac{8z}{17} , - \frac{2z}{17} , z \right) +$$

$$+ \sqrt{5^n}\begin{pmatrix} \mathrm{Cos}\, n\theta & \mathrm{Sin}\, n\theta & 0 \\ -\mathrm{Sin}\, n\theta & \mathrm{Cos}\, n\theta & 0 \\ 0 & 0 & 0 \end{pmatrix} \left(x- \frac{8z}{17},\ y + \frac{2z}{17}, 0 \right)$$

The trajectory goes to infinity while approaching the unstable subspace of M, unless the initial condition $\mathbf{x} = (x, y, z)$ is such that

$$x = \frac{8}{17} z \quad y = - \frac{2}{17} z .$$

In this case $\mathbf{x} \in \mathbf{W}_S$ and the orbit converges to $\mathbf{0}$.

Notice that by using the equality $M^n = TD^nT^{-1}$ we have found $M^n\mathbf{x} = M^n\mathbf{x}_S + M^n\mathbf{x}_U$ since

$$\mathbf{x} = (x, y, z) = \mathbf{x}_S + \mathbf{x}_U = \left(\frac{8z}{17}, \frac{-2z}{17}, z\right) + \left(x - \frac{8z}{17}, y + \frac{2z}{17}, 0\right).$$

It is important to remember that the trajectory of every initial state which does not belong to the stable subspace will go to infinity. It may take many iterations to detect this behavior, particularly if the $\mathbf{x}_S$ component of the starting point is very large and the $\mathbf{x}_U$ component is very small. In fact, in this case the trajectory will move towards the origin during the initial iterations. Eventually, the trajectory will get closer to the unstable subspace, and from then on its states will move away from the equilibrium at an exponential rate.

In the case when some of the eigenvalues are not semisimple, we can reach the same conclusions about the dynamic behavior of the system generated by M on the basis of the following facts .

Recall that given a real eigenvalue r_j and given an eigenvector $\mathbf{x}_j$ corresponding to it, the subspace $\mathrm{Span}\{\mathbf{x}_j\}$ is invariant under the action of M. Likewise, given a pair of complex conjugate eigenvalues $a + ib$, $a - ib$ $(b \neq 0)$ the subspace $\mathrm{Span}\{\mathbf{u}, \mathbf{v}\}$, where the linearly independent vectors **u**, **v** are solution of (3.3.14), is invariant under the action of M.

We can expand the previous statement into a more comprehensive result. To every real eigenvalue r_j corresponds an invariant subspace $\mathbf{X}_j$ of dimension equal to the algebraic multiplicity of r_j. More precisely, we have

$$\mathbf{X}_j = \text{generalized kernel of } r_jI - M.$$

In fact, let $p \geq 1$ be the first integer such that

$$\text{Ker}(r_jI - M)^p = \text{Ker}(r_jI - M)^{p+1}.$$

Recall that $\text{Ker}(r_jI - M)^p$ is the generalized kernel of $r_jI - M$.
Notice that $\mathbf{x} \in \text{Ker}(r_jI - M)^p$ if and only if $\mathbf{x} \in \text{Ker}(r_jI - M)^{p+1}$, i.e.

$$(r_jI - M)^p \mathbf{x} = \mathbf{0} \quad \text{if and only if} \quad (r_jI - M)^{p+1} \mathbf{x} = \mathbf{0} .$$

From the last equality we derive

$$(r_jI - M)^p (r_j\mathbf{x} - M\mathbf{x}) = \mathbf{0}.$$

Since $(r_jI - M)^p r_j\mathbf{x} = \mathbf{0}$, we obtain

$$(r_jI - M)^p M\mathbf{x} = \mathbf{0},$$

which implies

$$\mathbf{x}_1 = M\mathbf{x} \in \text{Ker}(r_jI - M)^p .$$

Thus $\mathbf{X}_j = \text{Ker}(r_jI - M)^p$ is invariant under the action of M. In particular, for every $\mathbf{x}_0 \in \mathbf{X}_j$ we have $O(\mathbf{x}_0) \subset \mathbf{X}_j$.

With a more elaborate argument it can be proved that to every pair of complex conjugate eigenvalues $a \pm \mathbf{i}b$ corresponds an invariant subspace $\mathbf{Y}_j$ of dimension equal twice the algebraic multiplicity of $a + \mathbf{i}b$.

Remark 3.3.2 insures that

$$|r_j| = \rho(M)_{X_j} = \lim_{n\to\infty} \sqrt[n]{\|M^n\|_{X_j}},$$

$$\sqrt{a^2+b^2} = \rho(M)_{Y_j} = \lim_{n\to\infty} \sqrt[n]{\|M^n\|_{Y_j}}.$$

Extending the results of section 2 we obtain that in these invariant subspaces the origin is a sink or a source with respect to the action of M. More precisely, **0** is a sink if $|r_j| < 1$ or $a^2 + b^2 < 1$, and **0** is a source if $|r_j| > 1$ or $a^2 + b^2 > 1$.

Example 4.3.5 Let

$$M = \begin{pmatrix} 2 & 0 & 1 \\ 0 & 0.5 & 2 \\ 0 & 0 & 0.5 \end{pmatrix}.$$

The eigenvalues of M are $r_1 = 2$ and $r_2 = r_3 = 0.5$. The first is simple, the second is not semisimple, since its algebraic multiplicity is 2, while its geometric multiplicity is 1.

The generalized kernel of .5I - M has dimension 2 and it is found by considering the powers of .5I - M. We have,

$$(.5I - M)^2 = \begin{pmatrix} 2.25 & 0 & 1.5 \\ 0 & 0 & 0 \\ 0 & 0 & 0 \end{pmatrix},$$

whose kernel is $\mathrm{Ker}(.5I - M)^2 = \mathbf{X} = \mathrm{Span}\,\{\mathbf{e}_2, \mathbf{v} = (2, 0, -3)\}$.

Hence $\mathrm{Ker}(.5I - M)^2$ has dimension 2 and we do not need to go any further.

Assume that a vector **x** is a linear combination of $\mathbf{e}_2$ and **v**, i.e. $\mathbf{x} = a\mathbf{e}_2 + b\mathbf{v}$. Then

$$M\mathbf{x} = aM\mathbf{e}_2 + bM\mathbf{v} = 0.5\, a\mathbf{e}_2 - 6\, b\, \mathbf{e}_2 + 0.5\, b\, \mathbf{v} = (.5a - 6b)\mathbf{e}_2 + .5b\mathbf{v}.$$

Thus $M\mathbf{x} \in \mathrm{Ker}(.5I - M)^2$ which shows that **X** is invariant under the action of M. The only eigenvalue of M in this two-dimensional subspace is $r_2 = 0.5$. Consequently, the origin is a sink for the dynamical system defined by M on **X**.

Remark 4.3.1 The ideas illustrated above allow us to say that the two groups of eigenvalues G_S and G_U can now be formed also when some eigenvalues are not semisimple.

A basis of the stable subspace $\mathbf{W}_S$ is given by those linearly independent vectors which span the invariant subspaces $\mathbf{X}_j$ and $\mathbf{Y}_j$ just described, and correspond to eigenvalues r_j, $a + ib$ such that $|r_j| < 1$ and $a^2 + b^2 < 1$. The subspace $\mathbf{W}_U$ is obtained in a similar manner. This time r_j and $a + ib$ are such that $|r_j| > 1$ and $a^2 + b^2 > 1$.

3. Comparing trajectories from different initial states

A final remark may be helpful. We have analyzed dynamical systems in $\mathbf{R}^q$ governed by matrices with some eigenvalues in modulus smaller than 1, and the others with modulus larger than 1. In these cases we can express every vector $\mathbf{x}$ of $\mathbf{R}^q$ as a sum $\mathbf{x} = \mathbf{x}_S + \mathbf{x}_U$, where $\mathbf{x}_S$ is a linear combination of the vectors spanning the stable subspace $\mathbf{W}_S$ while $\mathbf{x}_U$ is a linear combination of the vectors spanning the unstable subspace $\mathbf{W}_U$. The n-th state of the orbit of $\mathbf{x}$ is the sum of the n-th states of the orbits of $\mathbf{x}_S$ and $\mathbf{x}_U$.

Given two initial conditions $\mathbf{x}$ and $\mathbf{y}$, the distance between the corresponding states of $O(\mathbf{x})$ and $O(\mathbf{y})$ grows exponentially, unless the two initial states differ by a vector belonging to the stable subspace, i.e. $\mathbf{x} - \mathbf{y} \in \mathbf{W}_S$. In this case, the distance between corresponding states goes to zero.

Example 4.3.6 Let M be as in Example 4.3.1 and consider the initial states (1, 2) and (2, 1). Their distance is $\sqrt{2}$. Recall that (1,0), (4, 5) are the eigenvectors of M and observe that

$$(1, 2) = -\frac{3}{5}(1, 0) + \frac{2}{5}(4, 5), \quad (2, 1) = \frac{6}{5}(1, 0) + \frac{1}{5}(4, 5).$$

Therefore

$$M^n(1, 2) = M^n\left(-\frac{3}{5}(1, 0) + \frac{2}{5}(4, 5)\right) = -\frac{3}{5}2^{-n}(1, 0) + \frac{2}{5}3^n(4, 5),$$

and

$$M^n(2, 1) = M^n\left(\frac{6}{5}(1, 0) + \frac{1}{5}(4, 5)\right) = \frac{6}{5}2^{-n}(1, 0) + \frac{1}{5}3^n(4, 5).$$

For n very large we can neglect the contribution to the distance between $M^n(1, 2)$ and $M^n(2, 1)$ coming from the terms containing 2^{-n}.

Therefore, the distance between corresponding states is approximately given by the norm of the vector $\frac{1}{5}3^n(4, 5)$, which is $\frac{3^n}{5}\sqrt{41}$. The distance grows exponentially.

Let us now take the point (2, 2) = (1, 2) + (1, 0). Notice that (2, 2) - (1, 2) = (1, 0), a vector which belongs to the stable subspace of M.

From $(2, 2) = \frac{2}{5}(1, 0) + \frac{2}{5}(4, 5)$ we derive

$$M^n(2, 2) = \frac{2}{5}2^{-n}(1, 0) + \frac{2}{5}3^n(4, 5) .$$

We see that the distance between corresponding states of the two trajectories starting at the points $(1, 2)$ and $(2, 2)$ is given by the norm of the vector $2^{-n}(1, 0)$, which obviously goes to 0 as $n \to \infty$.

4. Affine systems

The information obtained on dynamical systems of the form

(4.3.3) $\qquad \mathbf{x}_{n+1} = M\mathbf{x}_n ,$

can be extended to affine systems

(4.3.4) $\qquad \mathbf{x}_{n+1} = M\mathbf{x}_n + \mathbf{b} = N(\mathbf{x}_n)$

by means of the conjugacy $B(\mathbf{x}) = \mathbf{x} + \mathbf{c}$, where $\mathbf{c}$ is the vector

$$\mathbf{c} = (I-M)^{-1}\mathbf{b}.$$

From the two equalities

(4.3.5) $\qquad N = BMB^{-1} \qquad M = TDT^{-1}$

we derive

(4.3.6) $\qquad N = BTDT^{-1}B^{-1} \qquad$ and $\qquad N^n = BTD^nT^{-1}B^{-1}.$

Equivalently,

(4.3.7) $\qquad N^nBT = BTD^n.$

In the case when $\rho(M) = \rho(D) < 1$ all trajectories of D converge to $\mathbf{0}$. Consequently all trajectories of N converge to the stationary state $\mathbf{c}$. In the case when all eigenvalues of M have modulus larger than 1, all non stationary trajectories of D will diverge to infinity. The same conclusion holds for the dynamical system (4.3.4).

Let us analyze the case when some eigenvalues of M have modulus larger than 1 and the others have modulus smaller than 1. From (4.3.7) we derive that the stable subspace $\mathbf{X}_S$ of D is changed into the stable subset $V_S = BT\mathbf{X}_S$ of N.

In fact, $\mathbf{x}_0 \in \mathbf{X}_S$ if and only if $\mathbf{y}_0 = BT\mathbf{x}_0 \in BT\mathbf{X}_S$. From (4.3.7),

$$D^n\mathbf{x}_0 \to \mathbf{0} \text{ and } BT(\mathbf{0}) = B(\mathbf{0}) = \mathbf{c},$$

we derive $N^n\mathbf{y}_0 \to \mathbf{c}$. Thus, $\mathbf{V}_S = BT\mathbf{X}_S$ is the stable invariant subset of N. Since $T\mathbf{X}_S = \mathbf{W}_S$ is the stable subspace of M we obtain

$$\text{(4.3.8)} \qquad \mathbf{V}_S = \mathbf{W}_S + \mathbf{c}.$$

Similarly, the unstable invariant subset of N is

$$\text{(4.3.9)} \qquad \mathbf{V}_U = \mathbf{W}_U + \mathbf{c}.$$

Both $\mathbf{V}_S$ and $\mathbf{V}_U$ are obtained as translations of subspaces and are called **linear varieties.**

Example 4.3.7 Let

$$M = \begin{pmatrix} 0.5 & 0 \\ 3 & 2 \end{pmatrix},$$

$\mathbf{b} = (1, -8)$. For the dynamical system $\mathbf{x}_{n+1} = N(\mathbf{x}_n) = M\mathbf{x}_n + (1, -8)$ we obtain $\mathbf{c} = (2, 2)$.

The eigenvalues of M are 0.5 and 2 with corresponding eigenvectors (1, -2) and (0, 1). Thus

$$D = \begin{pmatrix} 0.5 & 0 \\ 0 & 2 \end{pmatrix}, \qquad T = \begin{pmatrix} 1 & 0 \\ -2 & 1 \end{pmatrix}, \; T^{-1} = \begin{pmatrix} 1 & 0 \\ 2 & 1 \end{pmatrix}.$$

$\mathbf{W}_S = \text{span}\,\{(1, -2)\}$ is the stable subspace of M and $\mathbf{W}_U = \text{span}\{(0, 1)\}$ is the unstable subspace. Accordingly,

$$\mathbf{V}_S = \{(2 + u, 2 - 2u) : u \in \mathbf{R}\}, \; \mathbf{V}_U = \{(2, u + 2) : u \in \mathbf{R}\}$$

are the stable and unstable linear varieties of N. In fact, from $N^n = BTD^nT^{-1}B^{-1}$, and using $\mathbf{y}_0 \in \mathbf{V}_S$, $\mathbf{y}_0 = (2 + u_0, 2 - 2u_0)$, we obtain

$$N^n(\mathbf{y}_0) = \mathbf{y}_n = BTD^nT^{-1}B^{-1}(u_0 + 2, -2u_0 + 2) = BTD^nT^{-1}(u_0, -2u_0) = BTD^n(u_0, 0) =$$

$$= BT((\tfrac{1}{2})^n u_0, 0) = B((\tfrac{1}{2})^n u_0, -2(\tfrac{1}{2})^n u_0) = (2 + (\tfrac{1}{2})^n u_0, 2 - (\tfrac{1}{2})^{n-1} u_0).$$

The sequence $\{\mathbf{y}_n\}$ is contained in $\mathbf{V}_S$ and $\mathbf{y}_n \to (2, 2)$ as $n \to \infty$.

From $x = 2 + u$, $y = 2 - 2u$ we derive $2x + y = 6$. This equation represents the stable linear variety $\mathbf{V}_S$.

An orbit $O(\mathbf{x}_0) = O(x_0, y_0)$ has the form

$$\mathbf{x}_n = N^n(\mathbf{x}_0) = BTD^nT^{-1}B^{-1}(\mathbf{x}_0) = BTD^nT^{-1}(x_0 - 2, y_0 - 2) =$$

$$BT((.5)^n (x_0 - 2), 2^n(2x_0 + y_0 - 6)) =$$

$$= B((.5)^n (x_0 - 2), (.5)^n (-2x_0 + 4) + 2^n(2x_0 + y_0 - 6)) =$$

$$(2 + (.5)^n (x_0 - 2), 2 + (.5)^n (-2x_0 + 4) + 2^n(2x_0 + y_0 - 6)) .$$

The orbit converges to $(2, 2)$ if and only if $2x_0 + y_0 - 6 = 0$, i.e. $(x_0, y_0) \in \mathbf{V}_S$.

In the case when $2x_0 + y_0 - 6 \neq 0$, i.e. $(x_0, y_0) \notin \mathbf{V}_S$ the orbit goes to infinity while approaching $\mathbf{V}_U$.

Example 4.3.8 Let M be the matrix of Example 4.3.4 , and consider the dynamical system

$$\mathbf{x}_{n+1} = M\mathbf{x}_n + (2, -2, 2) = N\mathbf{x}_n.$$

We find that $\mathbf{c} = (1, -1, 4)$. The stable linear variety of N is

$$\mathbf{V}_S = \mathbf{W}_S + \mathbf{c} = \mathrm{Span}\{(8, -2, 17)\} + \mathbf{c} = \{(1 + 8u, -1 - 2u, 4 + 17u) : u \in \mathbf{R}\}.$$

The unstable linear variety of N is

$$\mathbf{V}_U = \mathbf{W}_U + \mathbf{c} = \mathrm{Span}\{\mathbf{e}_1, \mathbf{e}_2\} + \mathbf{c} = \{(1 + u, -1 + v, 4) : u,v \in \mathbf{R}\}.$$

Let us take an initial condition in $\mathbf{V}_S$. For example, with $u = 1$ we get $\mathbf{x}_0 = (9, -3, 21)$. Then

$$N^n(9, -3, 21) = BTD^nT^{-1}B^{-1}(9, -3, 21) = BTD^nT^{-1}(8, -2, 17) = BTD^n(1, 0, 0) =$$

$$= BT(\frac{1}{2^n}, 0, 0) = B(\frac{8}{2^n}, -\frac{2}{2^n}, \frac{17}{2^n}) = (1 + \frac{8}{2^n}, -1 - \frac{2}{2^n}, 4 + \frac{17}{2^n}) \to (1, -1, 4)$$

as $n \to \infty$.

Similarly, let us consider a point $\mathbf{x}_0 \in \mathbf{V}_U$. For example, $\mathbf{x}_0 = (0, 0, 4)$ corresponding to $u = -1$, $v = 1$. Then

$$N^n(0, 0, 4) = BTD^nT^{-1}B^{-1}(0, 0, 4) = BTD^nT^{-1}(-1, 1, 0) = BTD^n(0, -1, 1) =$$

$$= BT(\sqrt{5^n}(0, -\text{Cos}\, n\theta + \text{Sin}\, n\theta, \text{Cos}\, n\theta + \text{Sin}\, n\theta) =$$

$$= B(\sqrt{5^n}(-\text{Cos}\, n\theta + \text{Sin}\, n\theta, \text{Cos}\, n\theta + \text{Sin}\, n\theta, 0) =$$

$$= (1 + \sqrt{5^n}(-\text{Cos}\, n\theta + \text{Sin}\, n\theta),\ -1 + \sqrt{5^n}(\text{Cos}\, n\theta + \text{Sin}\, n\theta), 4).$$

We see that the orbit remains in $\mathbf{V}_U$ while going to ∞ as $n \to \infty$.

Section 4. Eigenvalues with Modulus 1

We now extend the analysis of the previous sections to include those cases when some of the eigenvalues have modulus 1. We again focus our attention on matrices with simple eigenvalues, or with eigenvalues having the same algebraic and geometric multiplicity (semisimple).

For organizational purposes we shall investigate separately the following situations:

- There is an eigenvalue equal 1.
- There is an eigenvalue equal -1.
- There is a pair of complex conjugate eigenvalues with modulus 1.

Before studying each single case we observe that the decomposition technique used in the analysis of a saddle stationary state allows us to decompose every vector of $\mathbf{R}^q$ into several parts, according to the eigenvalues of M.

For example, if 1 is a simple eigenvalue, we write each vector $\mathbf{x} \in \mathbf{R}^q$ as

$$\mathbf{x} = \mathbf{x}_1 + \mathbf{x}_R,$$

where $\mathbf{x}_1$ belongs to the invariant subspace spanned by an eigenvector corresponding to the eigenvalue 1 and $\mathbf{x}_R$ belongs to the invariant subspace corresponding to the remaining eigenvalues.

A similar decomposition can be used when

$$r = -1 \quad \text{or} \quad r = a \pm ib, \ a^2 + b^2 = 1,$$

are eigenvalues. In this way the action of M is decomposed into its action on suitable invariant subspaces, namely

- $\mathbf{X}_1$, corresponding to $r = 1$;
- $\mathbf{X}_{-1}$ corresponding to $r = -1$;
- $\mathbf{X}_C$ corresponding to $r = a \pm ib$, $a^2+b^2 = 1$;
- $\mathbf{X}_R$ corresponding to all eigenvalues r such that $|r| \neq 1$.

The action of M in this last subspace has been studied in the previous sections. Here we limit our analysis to the action of M on the first three.

1. The action of M on X_1

Assume $r_1 = 1$ is a simple eigenvalue. Let $\mathbf{u}_1$ be an eigenvector corresponding to it. From $M\mathbf{u}_1 = \mathbf{u}_1$ we derive that every element of $\mathbf{X}_1 = \text{span}\{\mathbf{u}_1\}$ is a stationary state of the dynamical system governed by M.

In the case when $r_1 = 1$ has algebraic and geometric multiplicity k, we will find k linearly independent eigenvectors corresponding to it, say $\mathbf{u}_1, \mathbf{u}_2, \ldots, \mathbf{u}_k$ and every element of the k-dimensional subspace $\mathbf{X}_1$ spanned by $\mathbf{u}_1, \mathbf{u}_2, \ldots, \mathbf{u}_k$ is a stationary state of the system governed by M.

Example 4.4.1 Let

$$M = \begin{pmatrix} 2^{-1} & 2 \\ 0 & 1 \end{pmatrix}.$$

The eigenvalues of M are 2^{-1} and 1. An eigenvector corresponding to the first eigenvalue is $\mathbf{u} = (1, 0)$. The eigenvector $\mathbf{v} = (4, 1)$ corresponds to 1. Every vector of $\mathbf{R}^2$ can be written as a linear combination of $\mathbf{u}$ and $\mathbf{v}$, $\mathbf{x} = a\mathbf{u} + b\mathbf{v}$. Hence

$$M^n\mathbf{x} = aM^n\mathbf{u} + bM^n\mathbf{v} = a2^{-n}\mathbf{u} + b\mathbf{v}.$$

We see that as $n \to \infty$ the trajectory starting from $\mathbf{x}$ converges to $b\mathbf{v}$. For example, for

$$\mathbf{x}_0 = (1, 1) = -3(1, 0) + (4, 1),$$

we have

$$\mathbf{x}_n = -\frac{3}{2^n}(1, 0) + (4, 1) \to (4, 1) \text{ as } n \to \infty.$$

Example 4.4.2 Let

$$M = \begin{pmatrix} 1 & 0 & 1 \\ 0 & 1 & 1 \\ 0 & 0 & .5 \end{pmatrix}.$$

The eigenvalues of M are $r_1 = r_2 = 1, r_3 = .5$. The second is simple, the first is semisimple with algebraic multiplicity 2. Two linearly independent eigenvectors corresponding to $r = 1$ are $\mathbf{e}_1$ and $\mathbf{e}_2$.

An eigenvector corresponding to $r_3 = .5$ is $\mathbf{z} = (-2, -2, 1)$.

Therefore, in the two-dimensional subspace $\mathbf{X}_1$ spanned by the two vectors $\mathbf{e}_1$ and $\mathbf{e}_2$ the action of M reduces to the identity, while in the one-dimensional subspace $\mathbf{X}_R$ spanned by $\mathbf{z}$ the action of M is a contraction of magnitude 0.5.
Every vector $\mathbf{x} \in \mathbf{R}^3$ can be written in a unique manner as $\mathbf{x} = a\mathbf{e}_1 + b\mathbf{e}_2 + c\mathbf{z}$. Thus,

$$\mathbf{x}_n = M^n\mathbf{x} = a\mathbf{e}_1 + b\mathbf{e}_2 + 2^{-n}c\mathbf{z} \to a\mathbf{e}_1 + b\mathbf{e}_2 \quad \text{as} \quad n \to \infty.$$

To see the importance of the assumption that the algebraic and geometric multiplicity be the same we now examine an example in which this condition is violated.

Example 4.4.3 Let

$$M = \begin{pmatrix} 1 & 1 & 0 \\ 0 & 1 & 1 \\ 0 & 0 & .5 \end{pmatrix}.$$

The eigenvalues of M are 1 and .5 . The second is simple, the first is not semisimple. Let $\mathbf{x}_0 = (1, 1, 0)$. Then $M\mathbf{x}_0 = (2, 1, 0)$ and in general,

$$M^n(1, 1, 0) = (n+1, 1, 0).$$

The orbit goes to infinity.

Therefore, when 1 is not semisimple we should expect the presence of unbounded orbits.

2. **The action of M on X_{-1}**

Assume that $r_1 = -1$ is a simple eigenvalue. Let $\mathbf{u}$ be an eigenvector corresponding to it. Then

$$M(\mathbf{u}) = -\mathbf{u} \quad \text{and} \quad M(M(\mathbf{u})) = M(-\mathbf{u}) = \mathbf{u}.$$

We see that the trajectory starting from every element (except $\mathbf{0}$) of X_{-1} is periodic of period 2.
In the case when $r_1 = -1$ has algebraic and geometric multiplicity k, we find k linearly independent eigenvectors $\mathbf{u}_1, \mathbf{u}_2, \ldots, \mathbf{u}_k$ corresponding to it. The trajectory starting at any element (except $\mathbf{0}$) of the k-dimensional subspace $\mathbf{X}_{-1}$ spanned by $\mathbf{u}_1, \mathbf{u}_2, \ldots, \mathbf{u}_k$ is periodic of prime period 2.

Example 4.4.4 Let

$$M = \begin{pmatrix} -1 & 1 \\ 0 & 2^{-1} \end{pmatrix}.$$

The eigenvalues of M are -1 and 2^{-1}. Eigenvectors corresponding to them are

$$\mathbf{u} = (1, 0) \text{ and } \mathbf{v} = (2, 3)$$

respectively. Every vector $\mathbf{y} \in \mathbf{R}^2$ can be written as a linear combination of $\mathbf{u}$ and $\mathbf{v}$, $\mathbf{y} = a\mathbf{u} + b\mathbf{v}$. Hence

$$M^n\mathbf{y} = aM^n\mathbf{u} + bM^n\mathbf{v} = a(-1)^n\mathbf{u} + b2^{-n}\mathbf{v}.$$

As $n \to \infty$ the trajectory starting from $\mathbf{y}$ converges to the periodic orbit of period 2,

$$\mathbf{x}_0 = a\mathbf{u}, \ \mathbf{x}_1 = -a\mathbf{u}.$$

For example, with

$$\mathbf{y} = (1, 1) = \frac{1}{3}(1, 0) + \frac{1}{3}(2, 3),$$

we have $\mathbf{y}_{2n} \to \frac{1}{3}\mathbf{u}$ and $\mathbf{y}_{2n+1} \to -\frac{1}{3}\mathbf{u}$ as $n \to \infty$.

When $r = -1$ is not semisimple we have the same problems outlined in Example 4.4.3 for the case $r = 1$.

3. **The action of M on X_C**

We consider the case when M has a pair of complex conjugate eigenvalues with modulus 1, say $r_1 = a + ib$ and $r_2 = a - ib$ with $\sqrt{a^2 + b^2} = 1$. Let us first assume that the two eigenvalues are simple. There is an invariant subspace X_C of dimension 2 spanned by two linearly independent vectors $\mathbf{u}$ and $\mathbf{v}$ found by solving the equations (see 3.3.14)

$$M\mathbf{u} = a\mathbf{u} - b\mathbf{v}, \qquad \text{and} \qquad M\mathbf{v} = b\mathbf{u} + a\mathbf{v}.$$

We can place the components of $\mathbf{u}$ and $\mathbf{v}$ in the first and second column of the matrix T.

The result of this move is $Te_1 = \mathbf{u}$ and $Te_2 = \mathbf{v}$. The subspace $\mathbf{Y}$ spanned by $\mathbf{e}_1$ and $\mathbf{e}_2$ is invariant under the action of the real canonical form D of M, $D = T^{-1}MT$.
Every vector $\mathbf{y} \in \mathbf{Y}$ can be written as $\mathbf{y} = (x, y, 0, ..., 0)$ and, according to the results of Section 3 of this chapter, we know that

$$D(x, y, 0, ...,0) = (x\mathrm{Cos}\,\theta + y\mathrm{Sin}\,\theta, -x\mathrm{Sin}\,\theta + y\mathrm{Cos}\,\theta, 0, ..., 0)\ .$$

The action of D on $\mathbf{Y}$ is a rotation of an angle θ, and

$$D^n(x, y, 0, ..., 0) = (x\mathrm{Cos}(n\theta) + y\mathrm{Sin}(n\theta), -x\mathrm{Sin}(n\theta) + y\mathrm{Cos}(n\theta), 0, ..., 0).$$

In the case when the ratio between θ and π is rational there exists a positive integer p such that every non stationary trajectory in $\mathbf{Y}$ of the dynamical system governed by D is periodic of prime period p. Consequently every non stationary orbit of M in $\mathbf{X}_C$ is periodic of the same period. The following example illustrates the situation.

Example 4.4.5 Let

$$M = \begin{pmatrix} \sqrt{2} & -\sqrt{2} \\ 1/\sqrt{2} & 0 \end{pmatrix}.$$

The eigenvalues of M are $\frac{1}{\sqrt{2}}(1+\mathbf{i})$ and $\frac{1}{\sqrt{2}}(1-\mathbf{i})$. A pair of vectors $\mathbf{u}$ and $\mathbf{v}$ is

$$\mathbf{u} = (2, 1) \text{ and } \mathbf{v} = (0,-1).$$

The matrices D, T , T^{-1} are

$$D = \begin{pmatrix} \frac{1}{\sqrt{2}} & \frac{1}{\sqrt{2}} \\ -\frac{1}{\sqrt{2}} & \frac{1}{\sqrt{2}} \end{pmatrix} = \begin{pmatrix} \cos\frac{\pi}{4} & \sin\frac{\pi}{4} \\ -\sin\frac{\pi}{4} & \cos\frac{\pi}{4} \end{pmatrix},$$

$$T = \begin{pmatrix} 2 & 0 \\ 1 & -1 \end{pmatrix}, \qquad T^{-1} = \begin{pmatrix} 0.5 & 0 \\ 0.5 & -1 \end{pmatrix}.$$

The action of D is a rotation of $\pi/4$ and every non stationary orbit of the dynamical system governed by D is periodic of period 8. Thus, the trajectories of the dynamical system governed by M are periodic of period 8.

To confirm this let us consider the point $\mathbf{x}_0 = (1, 0)$. Then, with $r = \sqrt{2}$, we obtain

$$\mathbf{x}_1 = M(\mathbf{x}_0) = (r, r^{-1}),\ \mathbf{x}_2 = M(\mathbf{x}_1) = (1, 1),\ \mathbf{x}_3 = M(\mathbf{x}_2) = (0, r^{-1}),$$

$$\mathbf{x}_4 = M(\mathbf{x}_3) = (-1, 0),\ \mathbf{x}_5 = M(\mathbf{x}_4) = -(r, r^{-1}),\ \mathbf{x}_6 = M(\mathbf{x}_5) = (-1, -1),$$

$$\mathbf{x}_7 = M(\mathbf{x}_6) = -(0, r^{-1}),\ \mathbf{x}_8 = M(\mathbf{x}_7) = (1, 0) = \mathbf{x}_0.$$

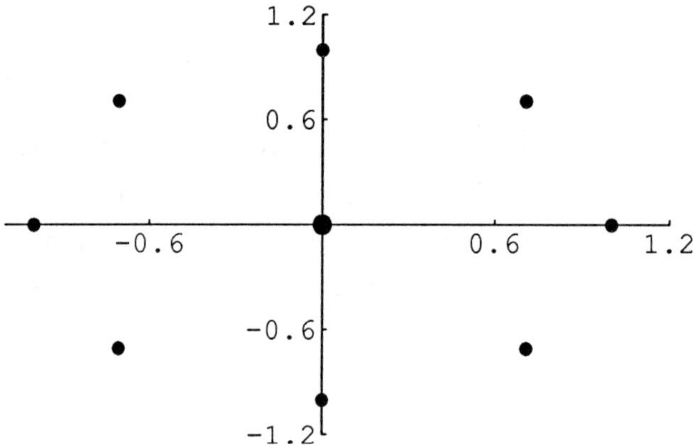

Fig. 4.4.1 The orbit of (1, 0) of the dynamical system governed by D.

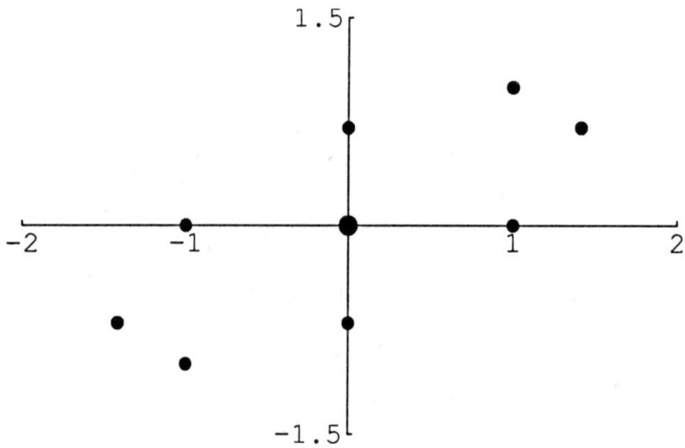

Fig. 4.4.2 The orbit of (1, 0) of the dynamical system governed by M.

Notice that the action of M (see Fig. 4.4.2) is not a rotation of $\theta = \pi/4$ or of a different angle since the angle α between $\mathbf{x}_1$ and $\mathbf{x}_2$ is such that

$$\cos\alpha = \frac{3}{\sqrt{10}},$$

while the angle between x_3 and x_4 is $\frac{\pi}{2}$.

Example 4.4.6 Let M be as in the previous example and consider the dynamical system

$$\mathbf{x}_{n+1} = M\mathbf{x}_n + (1, -1) = N(\mathbf{x}_n).$$

The stationary state of the system is $\mathbf{c} = (\frac{4+3r}{2}, \frac{2r+1}{2})$, $r = \sqrt{2}$. In the dynamical system governed by N all non stationary trajectories are periodic of period 8.

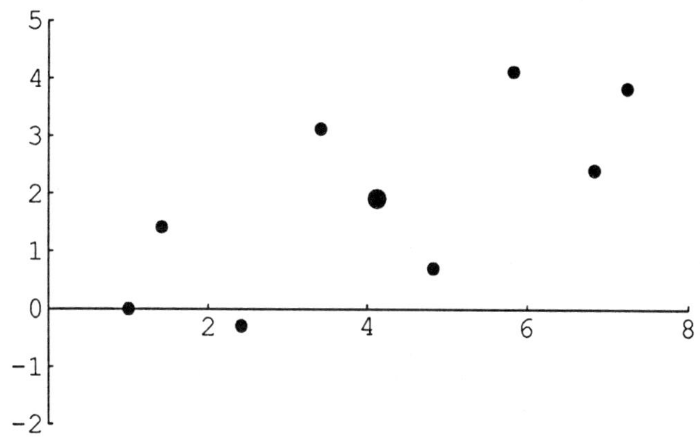

Fig. 4.4.3 The orbit starting at (1, 0) of the dynamical system governed by N.

The analysis can be extended to the case when $r_1 = a + ib$, $r_2 = a - ib$ are semisimple, but not simple. When the two eigenvalues are not semisimple we should expect the same kind of problems found before for the action on $\mathbf{X}_1$ and $\mathbf{X}_{-1}$.

4. **Conjugacy in the degenerate case**

When $r = 1$ is an eigenvalue of M it may not be possible to establish a conjugacy between the two systems

(4.4.1) $\qquad \mathbf{x}_{n+1} = M\mathbf{x}_n + \mathbf{b}$,

(4.4.2) $\qquad \mathbf{y}_{n+1} = M\mathbf{y}_n$.

The equation

$$\text{(4.4.3)} \qquad \mathbf{c} - M\mathbf{c} = \mathbf{b}$$

may have infinitely many solutions or it may have none. Since $\dim\mathrm{Ker}(I - M) \geq 1$ we obtain (see equality (3.3.7))

$$\text{(4.4.4)} \qquad \dim\mathrm{Im}(I - M) \leq q - 1 .$$

Consequently, (4.4.3) has infinitely many solutions when $\mathbf{b} \in \mathrm{Im}(I - M)$ and no solutions otherwise. We study both cases.

Assume that the eigenvalue $r = 1$ is semisimple. Denote by $\mathbf{X}_1$, $\mathbf{X}_R$ the invariant subspaces corresponding respectively to $r = 1$ and to the remaining eigenvalues of M. Every $\mathbf{x} \in \mathbf{R}^q$ can be written in a unique way as

$$\text{(4.4.5)} \qquad \mathbf{x} = \mathbf{x}_1 + \mathbf{x}_R, \quad \mathbf{x}_1 \in \mathbf{X}_1, \ \mathbf{x}_R \in \mathbf{X}_R .$$

The equation $\mathbf{c} - M\mathbf{c} = \mathbf{b}$ takes the form

$$\text{(4.4.6)} \qquad \mathbf{c}_1 + \mathbf{c}_R - M\mathbf{c}_1 - M\mathbf{c}_R = \mathbf{b}_1 + \mathbf{b}_R.$$

Since $M\mathbf{c}_1 = \mathbf{c}_1$ we obtain

$$\text{(4.4.7)} \qquad \mathbf{c}_R - M\mathbf{c}_R = \mathbf{b}_1 + \mathbf{b}_R.$$

There are two possibilities : $\mathbf{b}_1 = \mathbf{0}$, $\mathbf{b}_1 \neq \mathbf{0}$.

A. $\mathbf{b}_1 \neq \mathbf{0}$.

Equation (4.4.7) does not have any solution since $\mathbf{c}_R - M\mathbf{c}_R \in \mathbf{X}_R$, while $\mathbf{b}_1 \in \mathbf{X}_1$. No conjugacy can be found between the two systems (4.4.1) and (4.4.2). We need to study (4.4.1) directly.

Let $\mathbf{x}_0 = \mathbf{x}_{10} + \mathbf{x}_{R0}$. Since $M\mathbf{x}_{10} = \mathbf{x}_{10}$ and $M\mathbf{b}_1 = \mathbf{b}_1$ we obtain

$$\mathbf{x}_1 = \mathbf{x}_{10} + M\mathbf{x}_{R0} + \mathbf{b}_1 + \mathbf{b}_R ;$$

$$\mathbf{x}_2 = \mathbf{x}_{10} + M^2\mathbf{x}_{R0} + \mathbf{b}_1 + M\mathbf{b}_R + \mathbf{b}_1 + \mathbf{b}_R = \mathbf{x}_{10} + 2\mathbf{b}_1 + M^2\mathbf{x}_{R0} + M\mathbf{b}_R + \mathbf{b}_R ,$$

and in general,

$$\mathbf{x}_n = \mathbf{x}_{10} + n\mathbf{b}_1 + M^n\mathbf{x}_{R0} + M^{n-1}\mathbf{b}_R + \dots + M\mathbf{b}_R + \mathbf{b}_R .$$

The component of $\mathbf{x}_n$ in the subspace $\mathbf{X}_1$ goes to infinity. Consequently, every orbit of the system goes to infinity.

Example 4.4.7 Let

$$M = \begin{pmatrix} 1 & 1 \\ 0 & 0.5 \end{pmatrix}, \qquad \mathbf{b} = (3, 1)\,.$$

The eigenvalues of M are 1 and .5 with corresponding eigenvectors (1, 0) and (2, -1). Since

$$\mathbf{b} = 5(1, 0) - (2, -1)$$

we have $\mathbf{b}_1 \neq \mathbf{0}$. Hence, every orbit of the system

$$\mathbf{x}_{n+1} = M\mathbf{x}_n + \mathbf{b}$$

goes to infinity.

B. $\mathbf{b}_1 = \mathbf{0}$.

Equation (4.4.7) becomes

$$\text{(4.4.8)} \qquad \mathbf{c}_R - M\mathbf{c}_R = \mathbf{b}_R.$$

All solutions of (4.4.6) are obtained by adding to the unique solution $\mathbf{c}_R$ of (4.4.8) any vector of $\mathbf{X}_1$. Let $\mathbf{c}_0$ be one of the infinitely many solutions obtained,

$$\text{(4.4.9)} \qquad \mathbf{c}_0 = \mathbf{c}_{10} + \mathbf{c}_R, \qquad \mathbf{c}_{10} \in \mathbf{X}_1.$$

Set $B\mathbf{x} = \mathbf{x} + \mathbf{c}_0$. We now have a conjugacy between (4.4.1) and (4.4.2). Therefore, we can confine our analysis to (4.4.2) and use the conjugacy to derive conclusions about (4.4.1).

Assume, for example, that the origin is a sink for the action of M on $\mathbf{X}_R$. Then, given an initial condition $\mathbf{x}_0 = \mathbf{x}_{10} + \mathbf{x}_{R0}$, we have

$$\mathbf{x}_1 = M\mathbf{x}_0 = \mathbf{x}_{10} + M\mathbf{x}_{R0},$$
$$\mathbf{x}_2 = M\mathbf{x}_1 = \mathbf{x}_{10} + M^2\mathbf{x}_{R0}.$$

In general,

$$\mathbf{x}_n = \mathbf{x}_{10} + M^n\mathbf{x}_{R0}.$$

Since $M^n \mathbf{x}_{R0} \to \mathbf{0}$ as $n \to \infty$ we obtain $\mathbf{x}_n \to \mathbf{x}_{10}$. Moreover,

$$N^n \mathbf{x}_0 = BM^nB^{-1}\mathbf{x}_0 = BM^n(\mathbf{x}_0 - \mathbf{c}_0).$$

The sequence $\{M^n(\mathbf{x}_0 - \mathbf{c}_0)\}$ converges to $\mathbf{x}_{10} - \mathbf{c}_{10}$. From

$$B(\mathbf{x}_{10} - \mathbf{c}_{10}) = \mathbf{x}_{10} - \mathbf{c}_{10} + \mathbf{c}_0 = \mathbf{x}_{10} + \mathbf{c}_R,$$

we derive

$$(4.4.10) \qquad N^n \mathbf{x}_0 \to \mathbf{x}_{10} + \mathbf{c}_R.$$

Notice that the final result is independent of $\mathbf{c}_{10}$. Other cases are analyzed in a similar manner.

Example 4.4.8 Let $M = \begin{pmatrix} 1 & 0 \\ 1 & -0.5 \end{pmatrix}$, $\qquad \mathbf{b} = (0, 2)$.

The eigenvalues of M are 1 and -0.5. An eigenvector corresponding to 1 is $\mathbf{u} = (3, 2)$ and an eigenvector corresponding to -0.5 is $\mathbf{e}_2$. Consequently, $\mathbf{b} = \mathbf{b}_R = 2\mathbf{e}_2$ and $\mathbf{b}_1 = \mathbf{0}$. A solution of $\mathbf{c}_0 - M\mathbf{c}_0 = \mathbf{b}$, $\mathbf{c}_0 = (c_1, c_2)$ must satisfy the equation $-2c_1 + 3c_2 = 4$.
We could choose $c_1 = 1, c_2 = 2$. Then

$$\mathbf{c}_0 = (1, 2) = \frac{1}{3}(3, 2) + \frac{4}{3}(0, 1) = \mathbf{c}_{10} + \mathbf{c}_R, \quad \mathbf{c}_{10} = \frac{1}{3}(3, 2), \; \mathbf{c}_R = \frac{4}{3}(0, 1).$$

A different choice of $\mathbf{c}_0$ could have been $c_1 = c_2 = 4$. Then,

$$\mathbf{c}_0 = (4, 4) = \frac{4}{3}(3, 2) + \frac{4}{3}(0, 1) = \mathbf{c}_{10} + \mathbf{c}_R, \quad \mathbf{c}_{10} = \frac{4}{3}(3, 2), \; \mathbf{c}_R = \frac{4}{3}(0, 1).$$

The vector $\mathbf{c}_R$ does not change.

Notice that the origin is a sink for the action of M on $\mathbf{X}_R$. Consequently, given the initial condition

$$\mathbf{x}_0 = (2, 3) = \frac{2}{3}(3, 2) + \frac{5}{3}(0, 1) = \mathbf{x}_{10} + \mathbf{x}_R,$$

with

$$\mathbf{x}_{10} = (2, \frac{4}{3}), \; \mathbf{x}_R = (0, \frac{5}{3}),$$

we obtain that the orbit $O(\mathbf{x}_0)$ converges to $\mathbf{x}_{10} + \mathbf{c}_R = (2, \frac{4}{3}) + (0, \frac{4}{3}) = (2, \frac{8}{3})$.

Problems

Section 1.

4.1.1 Let

$$M = \begin{pmatrix} 1 & 2 \\ -1 & 3 \end{pmatrix}$$

and $\mathbf{b} = (1, 1)$. Write the first three steps of the trajectory of the dynamical system governed by the map $N(\mathbf{x}) = M(\mathbf{x}) + \mathbf{b}$, with initial condition $\mathbf{x}_0 = (0, 1)$.

4.1.2 Let M be as in the previous problem. Is the affine dynamical system defined in Problem 4.1.1 equivalent to the dynamical system governed by the map M ? In case of affirmative answer, find the vector $\mathbf{c}$ such that $\mathbf{c} - M\mathbf{c} = \mathbf{b}$. Is the vector $\mathbf{c}$ unique ?

4.1.3 Let M be as in Problem 4.1.1 and let $\mathbf{b}_1 = (2, -1)$. Consider the dynamical system governed by the map $N_1(\mathbf{x}) = M(\mathbf{x}) + \mathbf{b}_1$. Are N and N_1 conjugate? If the answer is affirmative you should find a conjugacy between the two systems.

4.1.4 Let A be a matrix similar to the matrix M of Problem 4.1.1. Consider the dynamical system governed by the map $T_2(\mathbf{x}) = A\mathbf{x} + \mathbf{b}$, where $\mathbf{b}$ is the same vector of Problem 4.1.1. Are T_2 and N conjugate? If the answer is affirmative you should find a conjugacy between the two systems.

4.1.5 Assume that the two systems $\mathbf{y}_{n+1} = M\mathbf{y}_n + \mathbf{b}$ and $\mathbf{x}_{n+1} = M\mathbf{x}_n$ are conjugate. Let $\mathbf{x}_0, \mathbf{x}_1, \mathbf{x}_2,$ be a periodic orbit of period 3 of $\mathbf{x}_{n+1} = M\mathbf{x}_n$. Find the corresponding periodic orbit of prime period 3 of $\mathbf{y}_{n+1} = M\mathbf{y}_n + \mathbf{b}$.

Section 2.

4.2.1 Let

$$M = \begin{pmatrix} 1 & -1 \\ 0.5 & -1 \end{pmatrix}.$$

Find two eigenvectors of M and a real canonical form D of M. Find a conjugacy between the dynamical systems governed by D and M. Derive the form of the trajectory $\mathbf{x}_{n+1} = M\mathbf{x}_n + (1, 1)$ with initial condition $\mathbf{x}_0 = (1, 2)$.

4.2.2 Let

$$M = \begin{pmatrix} 1 & 2 \\ -1 & 3 \end{pmatrix}.$$

Prove that every non stationary orbit of the system $\mathbf{x}_{n+1} = M\mathbf{x}_n$ diverges to infinity.

4.2.3 Let M be as in the previous problem. Find the stationary state of the dynamical system defined by $T(\mathbf{x}) = M\mathbf{x} + (1, 2)$.

4.2.4 Let

$$M = \begin{pmatrix} 2^{-1} & 2 \\ 0 & 3^{-1} \end{pmatrix}.$$

Prove that every orbit of the dynamical system defined by $T(\mathbf{x}) = M(\mathbf{x}) - (2, 5)$ is convergent. Find the stationary state of the system. Find the form of the orbit starting at $\mathbf{x}_0 = (2, -1)$.

4.2.5 Let

$$M = \begin{pmatrix} 2^{-1} & 0 & 1 \\ 0 & \frac{1}{2} & \frac{3}{4} \\ 0 & -\frac{3}{4} & \frac{1}{2} \end{pmatrix}.$$

Prove that every orbit of the dynamical system defined by M converges to $\mathbf{0}$. Find the form of the trajectory starting from $\mathbf{x}_0 = (1, 1, 2)$.

4.2.6 Let M be the matrix of the previous problem. Find M^{-1} and $\rho(M^{-1})$. Study the dynamical system $\mathbf{x}_{n+1} = M^{-1}\mathbf{x}_n + (2, 3)$.

Section 3.

4.3.1 Analyze the dynamical system defined in $\mathbf{R}^2$ by the matrix

$$M = \begin{pmatrix} 2^{-1} & 0 \\ 2 & 3 \end{pmatrix}.$$

In particular, find the eigenvalues and eigenvectors of M, and the stable and unstable subspace of M.

4.3.2 Let M be the matrix of Example 4.3.1 and consider the vector (1, 2). Notice that the trajectory starting from (1, 2) goes to infinity. Compare this trajectory with the one starting at the point (1.1, 2), and with the one starting at the point (1, 2.1). Make some comments on the dynamics of these three trajectories.

4.3.3 Let

$$M = \begin{pmatrix} 2^{-1} & 0 & 2 \\ 0 & 2 & 0 \\ 0 & 1 & 3 \end{pmatrix}.$$

Study the dynamical system in $\mathbf{R}^3$ governed by the matrix M.

4.3.4 Let

$$M = \begin{pmatrix} 2 & 0 & 1 \\ 0 & 2^{-1} & 4^{-1} \\ 0 & -1 & 2^{-1} \end{pmatrix}.$$

Verify that M has one real and two complex conjugate eigenvalues. Find an eigenvector corresponding to the real eigenvalue, and the two-dimensional invariant subspace corresponding to the pair of complex conjugate eigenvalues. Find the stable and unstable subspaces of M .

4.3.5 Let M be as in the previous example and let $\mathbf{b} = (1, 1, -1)$. Consider the dynamical system in $\mathbf{R}^3$ governed by the function $T(\mathbf{x}) = M(\mathbf{x}) + \mathbf{b}$. Find the stationary state of the system and the stable and unstable linear varieties of T .

4.3.6 Let M be as in Problem 4.3.1 and let $\mathbf{b} = (1, -2)$. Consider the dynamical system in $\mathbf{R}^2$ governed by the map $T(\mathbf{x}) = M^2(\mathbf{x}) + \mathbf{b}$. Find the stationary state of this system, and the stable and unstable linear varieties of T.

4.3.7 Study the system $\mathbf{x}_{n+1} = M\mathbf{x}_n + (1, 2)$ where M is the matrix

$$M = \begin{pmatrix} 2^{-1} & 2 \\ 0 & 2 \end{pmatrix}.$$

Find the stationary state and the stable and unstable linear varieties of the system.

4.3.8 Let $\mathbf{x}_{n+1} = M\mathbf{x}_n + (2, 3)$, with

$$M = \begin{pmatrix} 2^{-1} & 2 \\ 0 & 2 \end{pmatrix}.$$

Find the coordinates of $\mathbf{x}_{10}$ if $\mathbf{x}_0 = (3, 5)$.

Section 4.

4.4.1 Let

$$M = \begin{pmatrix} 1 & 0 \\ 2 & 2^{-1} \end{pmatrix}.$$

Find the eigenvalues and eigenvectors of M and study the dynamical system in $\mathbf{R}^2$ governed by M.

4.4.2 Let

$$M = \begin{pmatrix} 1 & 0 & 1 \\ 0 & 2^{-1} & 4^{-1} \\ 0 & -1 & 2^{-1} \end{pmatrix}.$$

Find the eigenvalues and eigenvectors of M and study the dynamical system in $\mathbf{R}^3$ governed by M.

4.4.3 Let

$$M = \begin{pmatrix} 1 & 0 & 1 \\ 0 & 1 & 0 \\ 0 & 0 & 2^{-1} \end{pmatrix}.$$

Find the eigenvalues and eigenvectors of M and study the dynamical system in $\mathbf{R}^3$ governed by M.

4.4.4 Let

$$M = \begin{pmatrix} 2 & 0 & 1 \\ 0 & 1 & 0 \\ 0 & 0 & 1 \end{pmatrix}.$$

Find the eigenvalues and eigenvectors of M and study the dynamical system in $\mathbf{R}^3$ governed by M.

4.4.5 Let

$$M = \begin{pmatrix} -1 & 0 \\ 1 & 1 \end{pmatrix}.$$

Find the eigenvalues and eigenvectors of M and study the dynamical system generated in $\mathbf{R}^2$ by the matrix M.

4.4.6 Let

$$M = \begin{pmatrix} -1 & 0 \\ 1 & -1 \end{pmatrix}.$$

Verify that -1 is the only eigenvalue of M and that its algebraic and geometric multiplicity are not the same. Study the dynamical system generated in $\mathbf{R}^2$ by the matrix M, using Example 4.4.3 as a guideline. Does the system have unbounded orbits ?

4.4.7 Let

$$M = \begin{pmatrix} \sqrt{2} & -\sqrt{2} \\ \frac{1}{\sqrt{2}} & 0 \end{pmatrix}$$

and let $\mathbf{b} = (1, 1)$.

Study the dynamical system in $\mathbf{R}^2$ governed by the map $N\mathbf{x} = M\mathbf{x} + \mathbf{b}$.

4.4.8 Let

$$M = \begin{pmatrix} 1 & 0 \\ 1 & \frac{1}{2} \end{pmatrix}$$

and let $\mathbf{b} = (1, 3)$. Study the dynamical system governed by $N\mathbf{x} = M\mathbf{x} + \mathbf{b}$.

4.4.9 Let

$$M = \begin{pmatrix} 1 & 0 \\ 1 & \frac{1}{2} \end{pmatrix}$$

and let $\mathbf{b} = (0, 2)$. Study the dynamical system governed by $N\mathbf{x} = M\mathbf{x} + \mathbf{b}$.

CHAPTER 5

NONLINEAR DYNAMICAL SYSTEMS

SUMMARY

The purpose of this chapter is to provide a qualitative analysis of stationary states and periodic orbits of nonlinear discrete dynamical systems. We first study the global stability of equilibrium points (section 1), followed by the local stability of equilibrium points and periodic orbits (section 2). We then turn our attention to repelling stationary states and periodic orbits (section 3) and to the presence of bounded invariant sets (section 4). In the last section we briefly discuss bifurcation.

Section 1. Global Stability of Equilibrium Points

Let $F : \mathbf{R}^q \to \mathbf{R}^q$ be continuous. To study the behavior of the dynamical system generated by F and to better understand what is happening in a particular but important situation, we need an extension of the theorem, usually presented in the first semester of Calculus, which states that a continuous function $f : [a, b] \to \mathbf{R}$ has a maximum and a minimum value.

The extension we need is that the interval [a, b] can be replaced by any closed ball in $\mathbf{R}^q$. The conclusion remains the same. Here is the result, stated more formally.

Lemma 5.1.1 *Let* D *be a closed ball in* $\mathbf{R}^q$ *and let* $f : D \to \mathbf{R}$ *be continuous. Then* f *has a maximum and a minimum value.*

1. Banach Contraction Principle

From the above result we derive a fundamental property of an important class of discrete dynamical systems in $\mathbf{R}^q$, namely the ones governed by the so-called contractive maps.

Definition of contraction

A map $F : \mathbf{R}^q \to \mathbf{R}^q$ is said to be a **contraction** if there exists a constant k, $0 \leq k < 1$, such that

$$(5.1.1) \qquad \|F(\mathbf{x}) - F(\mathbf{y})\| \leq k \, \|\mathbf{x} - \mathbf{y}\|$$

for all $\mathbf{x}, \mathbf{y} \in \mathbf{R}^q$.

Theorem 5.1.1 (Banach) *Let* $F : \mathbf{R}^q \to \mathbf{R}^q$ *be a* **contraction**. *Then* F *has one and only one equilibrium point* $\mathbf{x}_s$ *and for every* $\mathbf{x} \in \mathbf{R}^q$ *we have*

$$(5.1.2) \qquad \mathbf{x}_n = F(\mathbf{x}_{n-1}) \to \mathbf{x}_s \quad as \quad n \to \infty.$$

Proof. *Uniqueness*

The uniqueness of $\mathbf{x}_s$ is easy to establish, and it is left to the reader (Problem 5.1.1).

Convergence

The convergence of every trajectory to $\mathbf{x}_s$ is also easy to prove. In fact, let $\mathbf{x}_0$ be an arbitrary point of $\mathbf{R}^q$. Then

$$\|\mathbf{x}_1 - \mathbf{x}_s\| = \|F(\mathbf{x}_0) - F(\mathbf{x}_s)\| \leq k\,\|\mathbf{x}_0 - \mathbf{x}_s\|,$$

and in general,

$$\|\mathbf{x}_n - \mathbf{x}_s\| = \|F(\mathbf{x}_{n-1}) - F(\mathbf{x}_s)\| \leq k\|\mathbf{x}_{n-1} - \mathbf{x}_s\| \leq k^n\,\|\mathbf{x}_0 - \mathbf{x}_s\|.$$

Since $0 \leq k < 1$ we see that $\mathbf{x}_n = F(\mathbf{x}_{n-1}) \to \mathbf{x}_s$ as $n \to \infty$.

Existence

It remains to prove the existence of the equilibrium point. To this aim let $d = \|F(\mathbf{0})\|$ and choose a real number r such that $d \leq r(1 - k)$. The closed ball D centered at the origin and with radius r is mapped into itself by F.
In fact, for every $\mathbf{x} \in D$, we have $\|\mathbf{x}\| \leq r$ and

$$\|F(\mathbf{x})\| = \|F(\mathbf{x}) - F(\mathbf{0}) + F(\mathbf{0})\| \leq \|F(\mathbf{0})\| + \|F(\mathbf{x}) - F(\mathbf{0})\| \leq d + k\|\mathbf{x}\| \leq r(1-k) + kr = r.$$

Hence $F(\mathbf{x}) \in D$.
Let $\mathbf{x}_0$ be any point of D. With $\mathbf{x}_1 = F(\mathbf{x}_0)$, $\mathbf{x}_2 = F(\mathbf{x}_1)$, ..., we have

$$\|F(\mathbf{x}_1) - \mathbf{x}_1\| = \|F(\mathbf{x}_1) - F(\mathbf{x}_0)\| \leq k\,\|\mathbf{x}_1 - \mathbf{x}_0\| \leq k2r.$$

and in general,

$$\|F(\mathbf{x}_n) - \mathbf{x}_n\| = \|F(\mathbf{x}_n) - F(\mathbf{x}_{n-1})\| \leq k\|\mathbf{x}_n - \mathbf{x}_{n-1}\| \leq k^n 2r.$$

From the above inequalities and Lemma 5.1.1 we derive that the minimum value of the function $f : D \to \mathbf{R}$ defined by $f(\mathbf{x}) = \|F(\mathbf{x}) - \mathbf{x}\|$ is 0 and there is a point $\mathbf{x}_s \in D$, such that $f(\mathbf{x}_s) = 0$, i.e. $F(\mathbf{x}_s) = \mathbf{x}_s$.

Theorem 5.1.1 is known as **Banach Contraction Principle**. We should mention a variant of the BCP, due to the Italian mathematician **R. Caccioppoli**, who proved that the principle still holds if the role played by F in inequality (5.1.1) is played by one of its iterates. There are important cases in which F itself is not a contraction, but one of its iterates is. In these cases, Caccioppoli reminds us that the conclusions of the BCP remain unchanged.

Example 5.1.1 A more elaborate version of this example will be analyzed in Chapter 7 when we study Neural Networks. Let $h : \mathbf{R} \to \mathbf{R}$ be defined as follows

$$h(x) = \begin{cases} 0 & \text{if } x < 0 \\ x & \text{if } 0 \le x \le 1 \\ 1 & \text{if } x > 1 \end{cases}$$

and let M be the matrix

$$M = \begin{pmatrix} 0.2 & 0.5 \\ 0 & 0.4 \end{pmatrix}.$$

The operator norm of M, using the Euclidean norm, is not larger than 0.7. In fact $\|M(x, y)\|^2 = 0.04\,x^2 + 0.41\,y^2 + 0.2xy$, and this quantity should be maximized on the circle $x^2 + y^2 = 1$.

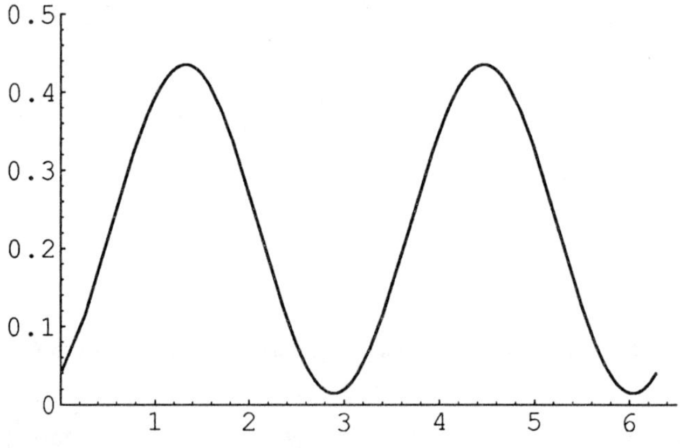

Fig. 5.1.1 The graph of $g(\theta)$ shows that the operator norm of M does not exceed 0.7.

We set $x = \text{Cos}\,\theta$ and $y = \text{Sin}\,\theta$, and find the maximum value of the function

$$g(\theta) = 0.04\,\text{Cos}^2\theta + 0.41\,\text{Sin}^2\theta + 0.2\,\text{Sin}\,\theta\,\text{Cos}\,\theta, \quad \theta \in [0, 2\pi].$$

The graph shows that the maximum value of $\|M(x, y)\|^2$ does not exceed 0.45. Hence the operator norm of M is not larger than 0.7.

Consequently, the function $F : \mathbf{R}^2 \to \mathbf{R}^2$ defined by

$$F(x, y) = M(h(x), h(y))$$

is a contraction. In fact, from $\|M\| \leq .7$ and $|h(x_1) - h(x_2)| \leq |x_1 - x_2|$, we derive

$$\|F(x, y) - F(u, v)\| \leq 0.7 \, \|(h(x), h(y)) - (h(u), h(v))\| \leq 0.7 \, \|(x, y) - (u, v)\|.$$

2. **Gradient maps**

There is another important case in which the dynamical situation is similar to the one envisioned by Theorem 5.1.1. We assume that F is a gradient. Namely, there exists a differentiable function $f : \mathbf{R}^q \to \mathbf{R}$ such that

$$(5.1.3) \qquad F(\mathbf{x}) = \nabla f(\mathbf{x}) = \left(\frac{\partial f}{\partial x_1}, \ldots, \frac{\partial f}{\partial x_q} \right).$$

We also assume f of class C^2, i.e. f is continuous together with its first and second partial derivatives. A result of Schwarz insures that

$$(5.1.4) \qquad \frac{\partial^2 f}{\partial x_i \partial x_j} = \frac{\partial^2 f}{\partial x_j \partial x_i}$$

for every $1 \leq i, j \leq n$. Hence the derivative $F_{\mathbf{x}}$ of F is a symmetric matrix at every point $\mathbf{x}$ of $\mathbf{R}^q$.

Example 5.1.2 Let

$$F(\mathbf{x}) = F(x, y) = (2x + y^3, 3xy^2 - 4y^3).$$

Then $F(x, y) = \nabla f(x, y)$ with $f(x, y) = x^2 + xy^3 - y^4$. Notice that

$$F_{\mathbf{x}}(\mathbf{x}) = \begin{pmatrix} 2 & 3y^2 \\ 3y^2 & 6xy\text{-}12y^2 \end{pmatrix}$$

is symmetric for every $\mathbf{x}$.

It is known that given a symmetric matrix M, the operator norm $\|M\|$, as defined in Chapter 3 and using the Euclidean norm in $\mathbf{R}^q$, satisfies the equality

(5.1.5) $\rho(M) = \|M\|.$

The spectral radius of the matrix equals its operator norm.

Besides equality (5.1.5) we need an important result, called Mean Value Inequality, which we shall use several times in this chapter.

Lemma 5.1.2 (Mean Value Inequality) *Let* $D = D(\mathbf{x}_0, r)$ *be a closed ball in* $\mathbf{R}^q$ *and let* $F : D \to \mathbf{R}^q$ *be continuous on* D. *Assume that* F *is differentiable on* $B(\mathbf{x}_0, r)$ *except possibly at finitely many points. Then, for every pair* $\mathbf{x}, \mathbf{y} \in D$ *there is a point* $\mathbf{c}$ *in the open line segment joining* $\mathbf{x}$ *with* $\mathbf{y}$, *such that*

(5.1.6) $\|F(\mathbf{x}) - F(\mathbf{y})\| \le \|F_{\mathbf{x}}(\mathbf{c})\| \, \|\mathbf{x} - \mathbf{y}\|.$

Recall that $\|F_{\mathbf{x}}(\mathbf{c})\|$ is the operator norm of the matrix $F_{\mathbf{x}}(\mathbf{c})$. Inequality (5.1.6) holds true in every norm, and it is the multi-dimensional version of Lemma 2.2.3 .
With this machinery set up we now obtain for discrete dynamical systems in $\mathbf{R}^q$ a result which is similar to the one established in Chapter 2 for $n = 1$ (see Theorem 2.3.2) .

Theorem 5.1.2 *Let* $F : \mathbf{R}^q \to \mathbf{R}^q$ *be a gradient. Assume that* F *has an equilibrium point* $\mathbf{x}_s$, *the derivative* $F_{\mathbf{x}}$ *is continuous everywhere, and*

(5.1.7) $\rho(F_{\mathbf{x}}(\mathbf{x})) < 1$ *for every* $\mathbf{x} \in \mathbf{R}^q$.

Then $\mathbf{x}_s$ *is the only equilibrium point of* F *and for every* $\mathbf{x}_0 \in \mathbf{R}^q$ *we have*

(5.1.8) $\mathbf{x}_n = F(\mathbf{x}_{n-1}) \to \mathbf{x}_s$ *as* $n \to \infty$.

Proof. *Uniqueness of the equilibrium point*

Assume that we have two distinct equilibria $\mathbf{y}_s$ and $\mathbf{x}_s$. From the Mean Value Inequality, (5.1.5) and (5.1.7) we obtain

$$\|\mathbf{x}_s - \mathbf{y}_s\| = \|F(\mathbf{x}_s) - F(\mathbf{y}_s)\| \le \|F_{\mathbf{x}}(\mathbf{c})\| \, \|\mathbf{x}_s - \mathbf{y}_s\| < \|\mathbf{x}_s - \mathbf{y}_s\|.$$

Dividing by $\|\mathbf{x}_s - \mathbf{y}_s\|$ we arrive at the inequality $1 < 1$. This contradiction shows that there can be at most one equilibrium point.

We now prove that every trajectory converges to it.

Let $\mathbf{x}_0$ be any point of $\mathbf{R}^q$ and let $r = \|\mathbf{x}_0\|$. By Lemma 5.1.1, equality (5.1.5), and assumption (5.1.7), the maximum value of the operator norm of $F_{\mathbf{x}}$, $\|F_{\mathbf{x}}(\mathbf{x})\|$, at any point $\mathbf{x}$ of the closed ball centered at $\mathbf{x}_s$ and with radius r, is smaller than or equal to a number $k < 1$. Therefore, using the Mean Value Inequality we find

$$\|\mathbf{x}_1 - \mathbf{x}_s\| = \|F(\mathbf{x}_0) - F(\mathbf{x}_s)\| \leq k\,\|\mathbf{x}_0 - \mathbf{x}_s\|$$

and in general , just as in Theorem 5.1.1,

$$\|\mathbf{x}_n - \mathbf{x}_s\| = \|F(\mathbf{x}_{n-1}) - F(\mathbf{x}_s)\| \leq k\,\|\mathbf{x}_{n-1} - \mathbf{x}_s\| \leq \ldots \leq k^n\|\mathbf{x}_0 - \mathbf{x}_s\| ,$$

which shows that the trajectory starting from $\mathbf{x}_0$ converges to $\mathbf{x}_s$.

As in the one-dimensional case, the assumption regarding the existence of an equilibrium point can be replaced by the assumption of the existence of a bounded orbit.

The reader has certainly recognized that Theorem 5.1.2 is not as sharp as its one-dimensional counterpart. The hypothesis that F is a gradient is automatically satisfied in that case. Hence we did not need to include it explicitly. The derivative was not assumed to be continuous, and its absolute value could equal 1 finitely many times in every bounded interval.

It would be useful to have a result in which the conclusion of Theorem 5.1.2 still holds, but the restriction that F be a gradient is removed. Unfortunately, at the present stage of research, such a result is not available.

3. Triangular maps

A map $F : \mathbf{R}^q \to \mathbf{R}^q$ is said to be **upper triangular** if

$$F(\mathbf{x}) = F(x_1, x_2, \ldots, x_q) = (F_1(x_1, \ldots, x_q), F_2(x_2, \ldots, x_q), \ldots, F_q(x_q)) ,$$

and **lower triangular** if

$$F(\mathbf{x}) = F(x_1, x_2, \ldots, x_q) = (F_1(x_1), F_2(x_1, x_2), \ldots, F_q(x_1, x_2, \ldots, x_q)).$$

The derivative of an upper triangular or a lower triangular map is respectively an upper or a lower triangular matrix.

Example 5.1.3 Let $F(x, y, z) = (x^2 + y + xz^2, yz, \text{Sin } z)$. Then F is upper triangular. In fact

$$F_1(x, y, z) = x^2 + y + xz^2, \quad F_2(y, z) = yz, \quad F_3(z) = \text{Sin } z.$$

Therefore

$$F_{\mathbf{x}}(\mathbf{x}) = \begin{pmatrix} 2x+z^2 & 1 & 2xz \\ 0 & z & y \\ 0 & 0 & \text{Cos } z \end{pmatrix},$$

and for $\mathbf{x}_0 = (1, 0, 0)$ we have

$$F_{\mathbf{x}}(1, 0, 0) = \begin{pmatrix} 2 & 1 & 0 \\ 0 & 0 & 0 \\ 0 & 0 & 1 \end{pmatrix}.$$

Example 5.1.4 Let $F(x, y, z) = (x^3, xy + \text{Sin } y, x + y + z^2)$. Then F is lower triangular.

For dynamical systems governed by such mappings we have the following important result which we present without proof.

Theorem 5.1.3 *Let* $F : \mathbf{R}^q \to \mathbf{R}^q$ *be an upper (or lower) triangular map. Assume that* $F(\mathbf{x}_s) = \mathbf{x}_s$ *for some* $\mathbf{x}_s \in \mathbf{R}^q$, $F_{\mathbf{x}}$ *is continuous and there is a constant* $k < 1$ *such that*

$$(5.1.9) \qquad \rho(F_{\mathbf{x}}(\mathbf{x})) \le k \quad \textit{for every } \mathbf{x} \in \mathbf{R}^q.$$

Then $\mathbf{x}_s$ *is the only equilibrium point of* F *and for every* $\mathbf{x}_0 \in \mathbf{R}^q$ *we have*

$$(5.1.10) \qquad \mathbf{x}_n = F(\mathbf{x}_{n-1}) \to \mathbf{x}_s \textit{ as } n \to \infty.$$

Because of the special form of the map F, assumption (5.1.9) is equivalent to

$$(5.1.11) \qquad \left|\frac{\partial F_i}{\partial x_i}(\mathbf{x})\right| \le k \quad \text{for all } \mathbf{x} \in \mathbf{R}^q \text{ and } i = 1, 2, ..., q.$$

In fact, since the matrix $F_{\mathbf{x}}(\mathbf{x})$ is a triangular matrix its eigenvalues are the entries of the main diagonal.

The reader may have noticed certain similarities between Theorems 5.1.2 and 5.1.3. In the one-dimensional case the two theorems coincide. In dimension higher than one the two results are different, since in Theorem 5.1.3 the derivative of F need not be a symmetric matrix.

Remark 5.1.1 Theorem 5.1.3 is true under less stringent assumptions. In particular, the q inequalities (5.1.11) can be replaced by the less restrictive inequalities

$$(5.1.12) \qquad \left|\frac{\partial F_i}{\partial x_i}(\mathbf{x})\right| \leq 1 \quad \text{for all } \mathbf{x} \in \mathbf{R}^q \text{ and } i = 1, 2, \ldots, q,$$

with the condition that for each $i = 1, 2, \ldots, q$ the equality in (5.1.12) holds only finitely many times in every closed ball.

Example 5.1.5 Let $F(\mathbf{x}) = F(x, y) = (y^2 + x/2, y/3)$. Every orbit of the dynamical system converges to $\mathbf{0}$ since $F(\mathbf{0}) = \mathbf{0}$ and $\rho(F_{\mathbf{x}}(\mathbf{x})) = 0.5$ for all $\mathbf{x} \in \mathbf{R}^2$. In fact

$$F_{\mathbf{x}}(\mathbf{x}) = \begin{pmatrix} 0.5 & 2y \\ 0 & 1/3 \end{pmatrix},$$

and the eigenvalues of $F_{\mathbf{x}}(\mathbf{x})$ are 0.5 and 1/3. Therefore, we can apply Theorem 5.1.3. Theorem 5.1.2 cannot be used since the map F is not a gradient.

The three previous results guarantee that all orbits converge to the unique stationary state of the system. Admittedly, their practical value is limited by the restrictions imposed on the function F. As a partial softening of this situation we should keep in mind that whenever F has the listed properties in an invariant closed ball D of $\mathbf{R}^q$, the results hold true for the dynamical system defined by F on D.

Section 2 . Sinks

The situations encountered in the previous section are certainly ideal, but the conditions required by them are quite restrictive. More frequently an equilibrium point is an attractor only for trajectories which start close to it.

We are now interested in finding conditions under which this behavior can be guaranteed. The result we prove is very similar to a companion result we obtained for the case of functions of one variable, and it involves again the derivative of the function at the equilibrium point. However, the presence of several variables makes the condition more difficult to verify. The proof is a bit more complex.

In Chapter 2 we mentioned that given a continuous function g, defined on an open interval I, and such that $|g(x_0)| < 1$ for some $x_0 \in I$ and given any real number k such that $|g(x_0)| < k < 1$, there exists an open interval J, $x_0 \in J$, with the property $|g(x)| \leq k$ for all $x \in J$.

We need an extension of this result in the same spirit as Lemma 5.1.1, namely with I and J replaced by open balls of $\mathbf{R}^q$. Formally stated, the result is as follows.

Lemma 5.2.1 *Let* $B = B(\mathbf{x}_0, r)$ *be an open ball of* $\mathbf{R}^q$ *and let* $g : B \to \mathbf{R}$ *be a continuous function. Assume that* $|g(\mathbf{x}_0)| < 1$.
Then for every real number k *which satisfies the inequality* $|g(\mathbf{x}_0)| < k < 1$, *there exists a real number* $d < r$ *with the property that* $|g(\mathbf{x})| \leq k < 1$ *whenever* $\mathbf{x} \in B(\mathbf{x}_0, d)$.

Some of the inequalities of Lemma 5.2.1 can be reversed, namely $|g(\mathbf{x}_0)| > 1$ implies that for every k with the property $|g(\mathbf{x}_0)| > k > 1$, there exists $d < r$ such that $|g(\mathbf{x})| \geq k > 1$ whenever $\mathbf{x} \in B(\mathbf{x}_0, d)$.

Moreover, both versions of the Lemma remain true in the case when g is not defined at $\mathbf{x}_0$, but

$$\lim_{\mathbf{x} \to \mathbf{x}_0} g(\mathbf{x}) = L \,, \quad |L| < 1 \ \text{ or } \ |L| > 1 \,.$$

Intuitively, we think of a sink as an equilibrium point $\mathbf{x}_s$ (or a periodic orbit $O(\mathbf{x}_0, p)$)

which attracts the orbit starting of every point $\mathbf{y}_0$ close by. We expect $O(\mathbf{y}_0)$ to remain close to $\mathbf{x}_s$ (the equilibrium point is stable) and the sequence of Euclidean distances

$$\{\|\mathbf{y}_n - \mathbf{x}_s\|: n = 0, 1, \ldots\}$$

to be decreasing and convergent to 0. The formal definition we propose insures this behavior (see Theorem 5.2.1). It is slightly different from the one presented in Chapter 2 for one-dimensional systems. The change is suggested by the goal of making it easier to verify that $\mathbf{x}_s$ (or $O(\mathbf{x}_0, p)$) is a sink.

Definition of Sink

Let $D = D(\mathbf{x}_s, r)$ and $F : D \to \mathbf{R}^q$ be continuous on D and C^1 on the open ball $B(\mathbf{x}_s, d)$. The stationary state $\mathbf{x}_s$ is said to be a **sink** or an **attractor** if $\rho(F_\mathbf{x}(\mathbf{x}_s)) < 1$.

Let $U \subset \mathbf{R}^q$ be open and $F : U \to \mathbf{R}^q$ be of class C^1. A periodic orbit of prime period p, $O(\mathbf{x}_0, p) = \{\mathbf{x}_0, \mathbf{x}_1, \ldots, \mathbf{x}_{p-1}, \mathbf{x}_p = \mathbf{x}_0, \mathbf{x}_1, \ldots\}$ is a **sink** or an **attractor** if each state $\mathbf{x}_i$ is an attracting stationary state of F^p.

Theorem 5.2.1 *Assume that* $\mathbf{x}_s$ *is a sink for* F. *Then there exist* $k < 1, d < r,$ *and* $T > 0$ *such that*

$$\|\mathbf{x}_n - \mathbf{x}_s\| \le Tk^n\|\mathbf{x}_0 - \mathbf{x}_s\|$$

whenever $\|\mathbf{x}_0 - \mathbf{x}_s\| \le d$.

Proof. Let $M = F_\mathbf{x}(\mathbf{x}_s)$. Since $\rho(M) < 1$, there exists a new norm $\|\,.\,\|_N$ in $\mathbf{R}^q$ such that the operator norm $\|M\|_N$ of M with respect to this new norm is smaller than 1.

Choose a number $k \in (\|M\|_N, 1)$ and let $g(\mathbf{x}) = \|F_\mathbf{x}(\mathbf{x})\|_N$. Since $F_\mathbf{x}$ is continuous, its operator norm with respect to $\|\,.\,\|_N$ is continuous as well. According to Lemma 5.2.1, there exists a positive number $d_0 < r$ such that for every $\mathbf{x}$, $\|\mathbf{x} - \mathbf{x}_s\|_N \le d_0$, we have $g(\mathbf{x}) \le k$. Using the Mean Value Inequality, we obtain that F maps the closed ball D_0 centered at $\mathbf{x}_s$ and with radius d_0 in the new norm, into itself and it is a contraction on D_0.

In fact, assuming $\|\mathbf{x} - \mathbf{x}_s\|_N \le d_0$ we obtain from the Mean Value Inequality

$$\|F(\mathbf{x}) - F(\mathbf{x}_s)\|_N \le \|F_\mathbf{x}(\mathbf{c})\|_N \, \|\mathbf{x} - \mathbf{x}_s\|_N ,$$

where $\mathbf{c}$ is some point in the open line segment joining $\mathbf{x}$ with $\mathbf{x}_s$.

Since $\|F_{\mathbf{x}}(\mathbf{c})\|_N \le k < 1$ we see that

$$\|F(\mathbf{x}) - \mathbf{x}_s\|_N = \|F(\mathbf{x}) - F(\mathbf{x}_s)\|_N \le k\, \|\mathbf{x} - \mathbf{x}_s\|_N.$$

Thus $F(\mathbf{x}) \in D_0$ for every $\mathbf{x} \in D_0$ and $F(\mathbf{x})$ is closer than $\mathbf{x}$ to the center $\mathbf{x}_s$ of D_0. Moreover, F is a contraction on D_0, since for every pair of points $\mathbf{x}, \mathbf{y} \in D_0$ we have

$$\|F(\mathbf{x}) - F(\mathbf{y})\|_N \le k\|\mathbf{x} - \mathbf{y}\|_N.$$

Using the same reasoning as in the proof of the Banach Contraction Principle we obtain that $\mathbf{x}_0 \in D_0$ implies

$$\|\mathbf{x}_n - \mathbf{x}_s\|_N \le k^n\|\mathbf{x}_0 - \mathbf{x}_s\|_N.$$

The inequality (see (4.2.4))

$$\|\mathbf{x}\| \le \|\mathbf{x}\|_N \le K\|\mathbf{x}\|$$

gives

$$\|\mathbf{x}_n - \mathbf{x}_s\| \le Kk^n\|\mathbf{x}_0 - \mathbf{x}_s\|.$$

whenever $\|\mathbf{x}_0 - \mathbf{x}_s\|_N \le d_0$. Choose $d < d_0$ so that $Kd < d_0$ and let $T = K$.

Example 5.2.1 Let $F(\mathbf{x}) = (x^2 - y, - y^3 + 5z , z/2 + x^2)$. Notice that $F(\mathbf{0}) = \mathbf{0}$ and $\rho(F_{\mathbf{x}}(\mathbf{0})) = \frac{1}{2}$. Hence $\mathbf{0}$ is an attractor for the dynamical system defined by F on $\mathbf{R}^3$. Not all trajectories of F converge to $\mathbf{0}$. For example, the orbit O((1, 1, 1)) is unbounded. Moreover, F has other equilibria (see Problem 5.2.1).

Notice that the orbits of every point $\mathbf{x}_0$ close to a sink $\mathbf{x}_s$ converge to $\mathbf{x}_s$ and $\|\mathbf{x}_n - \mathbf{x}_s\|$ goes to zero at an exponential rate.

Recall that the origin is a sink for the linear dynamical system generated by a matrix M if $\rho(M) < 1$, and that every trajectory converges to $\mathbf{0}$. The addition of nonlinear terms usually destroys this last global property. The local behavior, however, is preserved if the nonlinear part is small near the origin. More precisely, the following result holds.

Corollary 5.2.1 *Let* $F : \mathbf{R}^q \to \mathbf{R}^q$ *be of the form* $F(\mathbf{x}) = M\mathbf{x} + f(\mathbf{x})$. *Assume that*

- $\rho(M) < 1$;
- *there exist* $k > 0$, $a > 1$ *such that* $\|f(\mathbf{x})\| \le k\|\mathbf{x}\|^a$.

Then the origin is a sink.

Proof. Notice that $F(\mathbf{0}) = \mathbf{0}$ and $F_{\mathbf{x}}(\mathbf{0}) = M$, since

$$\lim_{\|\mathbf{x}\|\to 0} \frac{f(\mathbf{x})}{\|\mathbf{x}\|} = \mathbf{0}.$$

Apply Theorem 5.2.1 .

Example 5.2.2 Let

$$M = \begin{pmatrix} 0.5 & 3 \\ 0 & 0.25 \end{pmatrix},$$

and let $f(\mathbf{x}) = (x^2, xy)$. The origin is a stationary state of the dynamical system governed by the function $F(\mathbf{x}) = M\mathbf{x} + f(\mathbf{x})$. Moreover

$$\lim_{\|\mathbf{x}\|\to 0} \frac{(x^2, xy)}{\sqrt{x^2+y^2}} = \mathbf{0},$$

since

$$\frac{x^2}{\sqrt{x^2+y^2}} \le \sqrt{x^2+y^2} \text{ and } \frac{|xy|}{\sqrt{x^2+y^2}} \le \frac{1}{2} \sqrt{x^2+y^2} .$$

Hence Corollary 5.2.1 can be applied to the dynamical system governed by F. Since $\rho(M) = 0.5$, the origin is a sink.

We now establish for periodic orbits a result similar to Theorem 5.2.1. Its proof is based on a property of the spectrum for product of matrices.

Given the integers p-1, ..., 1, 0 we call **rotation** any rearrangement of the decreasing order p-1, ..., 2, 1, 0 of the form i-1, i-2, ..., 0, p-1, p-2, ..., i. For example, for

$p = 5$, the rearrangement 1, 0, 4, 3, 2 is a rotation, while 4, 3, 2, 0, 1 is not.

Proposition 5.2.1 *Let* $M_0, ..., M_{p-1}$ *be* p *matrices of size* q. *Then* $\text{Spec}(M_{p-1}...M_0)$ *remains unchanged when the order* p-1, ..., 1, 0 *is rearranged with a rotation.*

Proof. We shall confine ourselves to the case $p = 3$ and to the rotation 1, 0, 2. The reader can adjust this proof to the general case.

The determinants of $M_2M_1M_0$ and $M_1M_0M_2$ are equal. Hence 0 is in the spectrum of $M_2M_1M_0$ if and only if it is in the spectrum of $M_1M_0M_2$.
Now let $r \neq 0$ be a point in $\text{Spec}(M_2M_1M_0)$. For simplicity assume that r is real. Then there exists an eigenvector $\mathbf{x} \neq \mathbf{0}$, such that $M_2M_1M_0\mathbf{x} = r\mathbf{x}$. Notice that $M_1M_0\mathbf{x} \neq \mathbf{0}$. Apply M_1M_0 to both sides of $M_2M_1M_0\mathbf{x} = r\mathbf{x}$. We obtain

$$M_1M_0M_2M_1M_0\mathbf{x} = rM_1M_0\mathbf{x},$$

which shows that $M_1M_0\mathbf{x}$ is an eigenvector of $M_1M_0M_2$ corresponding to the same eigenvalue r.

We are now ready to establish a sufficient condition for a periodic orbit to be an attractor.

Theorem 5.2.2 *Let* $F : \mathbf{R}^q \to \mathbf{R}^q$ *be differentiable with a continuous derivative and let* $O(\mathbf{x}, p) = \{\mathbf{x}_0, \mathbf{x}_1, ..., \mathbf{x}_{p-1}, \mathbf{x}_p = \mathbf{x}_0, \mathbf{x}_1, ...\}$ *be a periodic orbit of prime period* $p > 1$. *Let* $M_0 = F^p{}_{\mathbf{x}}(\mathbf{x}_0)$ *and assume that* $\rho(M_0) < 1$. *Then* $O(\mathbf{x}_0, p)$ *is an attractor.*

Proof. The point $\mathbf{x}_0$ is a sink for the dynamical system defined by F^p.
Let $M_i = F^p{}_{\mathbf{x}}(\mathbf{x}_i)$, $i = 1, 2, ..., p - 1$. By the Chain Rule we have

$$M_i = F^p{}_{\mathbf{x}}(\mathbf{x}_i) = F_{\mathbf{x}}(\mathbf{x}_{i-1})\, F_{\mathbf{x}}(\mathbf{x}_{i-2})...F_{\mathbf{x}}(\mathbf{x}_0)F_{\mathbf{x}}(\mathbf{x}_{p-1})...F_{\mathbf{x}}(\mathbf{x}_i)\,.$$

This order corresponds to the rotation $i - 1, i - 2, ..., 2, 1, 0, p - 1, p - 2, ..., i + 1, i$ of the decreasing order $p - 1, p - 2, ..., i + 1, i, ..., 1, 0$ which represents the derivative of F^p at the point $\mathbf{x}_0$ (since $F^p{}_{\mathbf{x}}(\mathbf{x}_0) = F_{\mathbf{x}}(\mathbf{x}_{p-1})...F_{\mathbf{x}}(\mathbf{x}_0)$).
Hence, by Proposition 5.2.1, we have $\rho(M_i) < 1$, $i = 0, 1, ..., p - 1$.
By Theorem 5.2.1 each $\mathbf{x}_i$, $i = 0, 1, ..., p - 1$ is a sink for the dynamical system defined by F^p. Thus, the periodic orbit $O(\mathbf{x}_0, p)$ is an attractor.

Example 5.2.3 Let $F(x, y) = (1 - x^2 + y,\ y^2)$. Then $\{(0, 0), (1, 0)\}$ is a periodic orbit of period 2. We have

$$F_{\mathbf{x}}(\mathbf{0}) = \begin{pmatrix} 0 & 1 \\ 0 & 0 \end{pmatrix}, \quad F_{\mathbf{x}}(1, 0) = \begin{pmatrix} -2 & 1 \\ 0 & 0 \end{pmatrix}.$$

Therefore

$$F^2{}_{\mathbf{x}}(0, 0) = \begin{pmatrix} 0 & -2 \\ 0 & 0 \end{pmatrix},$$

and $\rho(F^2{}_{\mathbf{x}})(0, 0) = 0$. The periodic orbit is an attractor. Notice that

$$F^2{}_{\mathbf{x}}(1, 0) = \begin{pmatrix} 0 & 0 \\ 0 & 0 \end{pmatrix}.$$

The two matrices are different, but their spectra are the same.

In the first chapter we introduced the definition of a **stable stationary state** and a **stable periodic orbit**. Sufficient conditions for the stability of such orbits can be provided.

For example, if the point $\mathbf{x}_s$ is a stationary state and if for some $r > 0$ there is a ball $B = B(\mathbf{x}_s, r)$ such that $\|F_{\mathbf{x}}(\mathbf{x})\| \leq 1$ for all $\mathbf{x} \in B(\mathbf{x}_s, r)$, then $\mathbf{x}_s$ is stable. This result is the q-dimensional version of Theorem 2.2.5. The two proofs are identical.

Similarly, if $O(\mathbf{x}_0, p)$ is a periodic orbit of prime period p, and there exists $r > 0$ such that for every state $\mathbf{x}_k \in O(\mathbf{x}_0, p)$ we have $\|F_{\mathbf{x}}(\mathbf{x})\| \leq 1$ for all $\mathbf{x} \in B_k = B(\mathbf{x}_k, r)$, then $O(\mathbf{x}_0, p)$ is stable.

As in the case of one-dimensional systems (see Example 2.2.10 and also Example 5.3.4 below) it may happen that $\rho(F_{\mathbf{x}}(\mathbf{x}_i)) > 1$ for some of the indices $i = 0, 1, ..., p$ even though $\rho(M_0) < 1$ (and hence the same is true for all remaining indices).

Section 3. Repellers and Saddles

1. Repelling states

Definition of Source

As in the case of one-dimensional systems, we expect a repelling stationary state $\mathbf{x}_s$ (or periodic orbit $O(\mathbf{x}_0, p)$) to be unstable and to have the additional property that all orbits starting at points close to $\mathbf{x}_s$ will move away from it.

The technical definition we present insures this behavior (see Theorem 5.3.1). It is slightly different from the definition we proposed in the one-dimensional case. The change has the purpose of making it easier to verify that $\mathbf{x}_s$ is a source.

Let $D = D(\mathbf{x}_s, r)$ and $F : D \to \mathbf{R}^q$ be continuous on D and differentiable on the ball $B(\mathbf{x}_s, r)$, with continuous derivative. The stationary state $\mathbf{x}_s$ is said to be a **source** or a **repeller** if all eigenvalues of $F_{\mathbf{x}}(\mathbf{x}_s)$ have modulus larger than 1.

Let $U \subset \mathbf{R}^q$ be open and $F : U \to \mathbf{R}^q$ be of class C^1. A periodic orbit of prime period p, $O(\mathbf{x}_0, p) = \{\mathbf{x}_0, \mathbf{x}_1, ..., \mathbf{x}_{p-1}, \mathbf{x}_p = \mathbf{x}_0, \mathbf{x}_1, ...\}$ is a **source** or a **repeller** if each state $\mathbf{x}_i$ is a repelling stationary state of F^p.

Example 5.3.1 Let $F(x, y) = (x^2 + 2y, y^2 + 3x)$. The point $(0, 0)$ is a stationary state. Moreover, a simple computation shows that the two eigenvalues of the derivative of F at $(0, 0)$ have modulus larger than 1. Hence $(0, 0)$ is a repeller.

Example 5.3.2 Let $F(x, y) = (2x + y^2 - 2, x^2 - x - y^3)$. Then

$$F(1, 1) = (1, -1) \quad \text{and} \quad F(1, -1) = (1, 1).$$

Hence, we have a periodic orbit of period 2. The two eigenvalues of $F^2_{\mathbf{x}}(1,1)$ have modulus larger than 1. The periodic orbit is a source (see Problem 5.3.2).

In the following theorem we use several results previously established.

Theorem 5.3.1 *Let* $F : \mathbf{R}^q \to \mathbf{R}^q$ *be a dynamical system and assume that* F *is differentiable with continuous derivative. Let* $\mathbf{x}_s$ *be an equilibrium point of the systems and assume that* $\mathbf{x}_s$ *is a source. Then there exist* $d < r$, $k > 1$ *and* $T > 0$ *such that*

(5.3.1) $\mathbf{x}_0, \ldots, \mathbf{x}_{n-1} \in D(\mathbf{x}_s, d)$ *implies* $\|\mathbf{x}_n - \mathbf{x}_s\| \geq Tk^n\|\mathbf{x}_0 - \mathbf{x}_s\|$.

Proof. The eigenvalues of the matrix $M = F_{\mathbf{x}}(\mathbf{x}_s)$ have modulus larger than 1. Hence M is invertible and we can define a new norm in $\mathbf{R}^q$, $\|\,.\,\|_N$, such that $\|M^{-1}\|_N < 1$. Let $\mathbf{x}$, $\mathbf{y}$ be a pair of vectors . Then

(5.3.2) $$\|\mathbf{x} - \mathbf{y}\|_N = \|M^{-1}M\mathbf{x} - M^{-1}M\mathbf{y}\|_N \leq \|M^{-1}\|_N \, \|M\mathbf{x} - M\mathbf{y}\|_N ,$$

which shows that

(5.3.3) $$\|M\mathbf{x} - M\mathbf{y}\|_N \geq L \, \|\mathbf{x} - \mathbf{y}\|_N$$

with $L > 1$ (L is the reciprocal of $\|M^{-1}\|_N$). Since F is differentiable we have

(5.3.4) $$F(\mathbf{x}) - F(\mathbf{x}_s) = F_{\mathbf{x}}(\mathbf{x}_s)\,(\mathbf{x} - \mathbf{x}_s) + o(\|\mathbf{x} - \mathbf{x}_s\|_N) ,$$

where the higher order terms $o(\|\mathbf{x} - \mathbf{x}_s\|_N)$ have the property

$$\lim_{\|\mathbf{x} - \mathbf{x}_s\|_N \to 0} \frac{o(\|\mathbf{x} - \mathbf{x}_s\|_N)}{\|\mathbf{x} - \mathbf{x}_s\|_N} = \mathbf{0} .$$

From (5.3.4) we derive

$$\frac{\|F(\mathbf{x}) - F(\mathbf{x}_s)\|_N}{\|\mathbf{x} - \mathbf{x}_s\|_N} \geq \frac{\|F_{\mathbf{x}}(\mathbf{x}_s)(\mathbf{x} - \mathbf{x}_s)\|_N}{\|\mathbf{x} - \mathbf{x}_s\|_N} - \frac{\|o(\|\mathbf{x} - \mathbf{x}_s\|_N)\|_N}{\|\mathbf{x} - \mathbf{x}_s\|_N} \geq L - \frac{\|o(\|\mathbf{x} - \mathbf{x}_s\|_N)\|_N}{\|\mathbf{x} - \mathbf{x}_s\|_N}$$

which shows that

$$\lim_{\|\mathbf{x} - \mathbf{x}_s\|_N \to 0} \frac{\|F(\mathbf{x}) - F(\mathbf{x}_s)\|_N}{\|\mathbf{x} - \mathbf{x}_s\|_N} \geq L > 1 .$$

Hence, by the remark mentioned at the end of Lemma 5.2.1, and using as g the function

$$g(\mathbf{x}) = \frac{\|F(\mathbf{x}) - F(\mathbf{x}_s)\|_N}{\|\mathbf{x} - \mathbf{x}_s\|_N} ,$$

we obtain that for every k, $L > k > 1$ there exists $d_0 < r$ such that

(5.3.5) $$\|F(\mathbf{x}) - \mathbf{x}_s\|_N \geq k \, \|\mathbf{x} - \mathbf{x}_s\|_N$$

whenever $\|\mathbf{x} - \mathbf{x}_s\|_N \leq d_0$.

Consequently, as long as $\mathbf{x}_0, \ldots, \mathbf{x}_{n-1}$ belong to the closed ball centered at $\mathbf{x}_s$ and with radius d_0 in the norm $\|\,.\,\|_N$, we can use (5.3.5) repeatedly to obtain

$$(5.3.6) \qquad \|\mathbf{x}_n - \mathbf{x}_s\|_N = \|F(\mathbf{x}_{n-1}) - \mathbf{x}_s\|_N \geq k\,\|\mathbf{x}_{n-1} - \mathbf{x}_s\|_N \geq \ldots \geq k^n\|\mathbf{x}_0 - \mathbf{x}_s\|_N .$$

Hence, inequality (5.3.1) is true in the norm $\|\,.\,\|_N$ with $T = 1$. From (see (4.2.4))

$$\|\mathbf{x}\| \leq \|\mathbf{x}\|_N \leq K\|\mathbf{x}\| \qquad (K > 1),$$

we derive

$$\|\mathbf{x}_n - \mathbf{x}_s\| \geq K^{-1}k^n\|\mathbf{x}_0 - \mathbf{x}_s\|$$

whenever $\|\mathbf{x}_i - \mathbf{x}_s\|_N \leq d_0$, $i = 0, 1, \ldots, n\text{-}1$.
Choose $d < d_0$ so that $Kd < d_0$ and $T = K^{-1}$.

It is evident that as long as the orbit does not leave the closed ball $D = D(\mathbf{x}_s, d)$, the distance from the stationary state grows exponentially. Therefore the orbit must exit D. It may come back to it later, and it may even hit the stationary state $\mathbf{x}_s$.

Example 5.3.3 Let $F(x, y) = (x^2 + 2y, y^2 + 3x)$. The origin is a source.

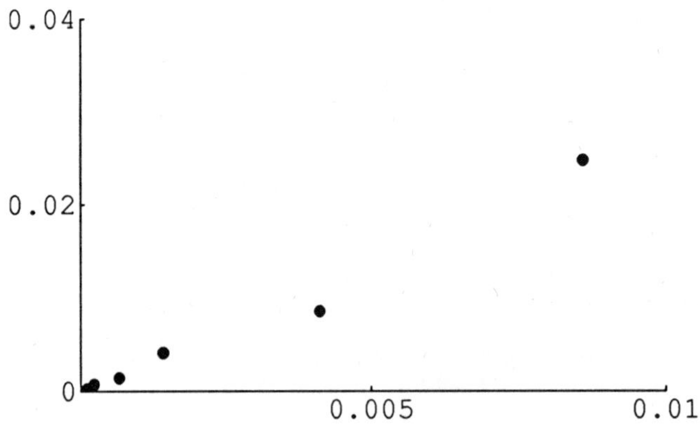

Fig. 5.3.1 By plotting ($\|\mathbf{x}_{i-1}\|$, $\|\mathbf{x}_i\|$) we can detect the exponential growth of the distances.

In Fig. 5.3.1 we have plotted the pairs $(\|\mathbf{x}_{i-1}\|, \|\mathbf{x}_i\|)$ for $i = 1, \ldots, 8$ starting at a point $\mathbf{x}_0$ very close to $\mathbf{0}$. Notice that the pairs of points seem to be close to the line $v = ru$ in the uv plane, with $r \approx 2.5$. This suggests the exponential growth of the distances since from $\|x_i\| \approx r\|x_{i-1}\|$ we derive

$$\|x_2\| \approx r^2\|x_0\|,\ \|x_3\| \approx r^3\|x_0\|, \ldots, \|x_n\| \approx r^n\|x_0\|.$$

The value $r \approx 2.5$ is expected since the spectral radius of the derivative of F at $\mathbf{0}$ gives us an initial good estimate of how fast the distances are increasing and $\rho(F_x(\mathbf{0})) = \sqrt{6}$.

Let $O(x_0, p)$ be a periodic source. By Theorem 5.3.1 there exist $d_0 < r_0$, $k_0 > 1$ and $T_0 > 0$ such that

$$(5.3.7) \qquad y_0, \ldots, y_{(n-1)p} \in D(x_0, d_0) = D_0$$

implies

$$\|y_{np} - x_0\| \geq Tk^n\|y_0 - x_0\| .$$

In other words, as long as the states $y_0, y_{2p}, \ldots, y_{(n-1)p}$ remain in the closed ball D_0 their distance from x_0 grows exponentially. A similar conclusion holds at every other state x_i of the periodic orbit.

In the case of dynamical systems in several dimensions we find a situation similar to the one mentioned for dynamical systems in the real line.
Given a repelling periodic orbit of prime period p, $O(x_0, p)$, and an initial state y_0 close to x_0, the distance $\|y_n - x_n\|$ may not grow with n, but only with np, i.e. over an entire period. The following example illustrates this situation.

Example 5.3.4 Let

$$F(x, y) = (-\frac{5}{3}x^2 + \frac{2}{3}y + 1, -\frac{5}{3}y^2 + \frac{2}{3}x + 1).$$

The orbit $\{x_0 = (0, 0), x_1 = (1, 1)\}$ is periodic of period 2.
The eigenvalues of the derivative of F at the origin are $r_1 = -r_2 = -2/3$, while the eigenvalues at (1, 1) are $s_1 = -4$ and $s_2 = -8/3$. The eigenvalues of the derivative of $F(F(\mathbf{x}))$ at the origin are $t_1 = 8/3$, $t_2 = -16/9$. Hence the orbit is a source.

From $\mathbf{y}_0 = (.01, .01)$ we get $\mathbf{y}_1 = (1.0065, 1.0065)$, $\mathbf{y}_2 \approx (-0.0174, -0.0174)$. We see that $\|\mathbf{y}_1 - \mathbf{x}_1\| < \|\mathbf{y}_0 - \mathbf{x}_0\|$, but $\|\mathbf{y}_2 - \mathbf{x}_0\| \approx \frac{16}{9} \|\mathbf{y}_0 - \mathbf{x}_0\|$. The separation constant is approximately equal to the smallest eigenvalue of the derivative of $F(F(\mathbf{x}))$ at the origin.

2. Saddles

So far we have examined the behavior of a dynamical system $F : \mathbf{R}^q \to \mathbf{R}^q$ around an equilibrium state or a periodic orbit in the case when all eigenvalues of the derivative have modulus smaller than 1, or larger than 1. Following the guidelines of the linear situation, we now analyze the so-called saddle case. We shall confine ourselves to equilibrium points.

Definition of saddle

A stationary state $\mathbf{x}_s$ is said to be a **saddle** when some eigenvalues of $F_\mathbf{x}(\mathbf{x}_s)$ have modulus smaller than 1, and the others have modulus larger than 1.

In the linear case, we found an invariant subspace $\mathbf{W}_S$ corresponding to the group of eigenvalues with modulus smaller than 1, and another $\mathbf{W}_U$ corresponding to the eigenvalues with modulus larger than 1. They were called **stable** and **unstable subspaces** of the system. In the nonlinear case the situation is more complex .

Assume for simplicity that $\mathbf{x}_s = \mathbf{0}$ and F is invertible. Denote by G the inverse of F .

Find $\mathbf{W}_S$ and $\mathbf{W}_U$ for $F_\mathbf{x}(\mathbf{0})$. These two subspaces constitute the first order approximation of $\mathbf{M}_S$ and $\mathbf{M}_U$, which are now called the **stable** and **unstable manifolds** of F at $\mathbf{0}$. The stable manifold $\mathbf{M}_S$ is defined as the set of those initial states $\mathbf{x}_0$ such that $F^n(\mathbf{x}_0) \to \mathbf{0}$ as $n \to \infty$, while the unstable manifold $\mathbf{M}_U$ is defined as the set of those initial states $\mathbf{x}_0$ such that $G^n(\mathbf{x}_0) \to \mathbf{0}$ as $n \to \infty$.

Example 5.3.5 Let $F(x, y) = (2x - y^3, 0.5\, y)$. Notice that $F(0, 0) = (0, 0)$ is the only equilibrium point of F . The inverse of F is the map

$$G(u, v) = (\frac{u + 8v^3}{2}, 2v) .$$

The derivative of F at (0, 0) is the matrix

$$F_x(\mathbf{0}) = \begin{pmatrix} 2 & 0 \\ 0 & 0.5 \end{pmatrix}.$$

The stable subspace of this matrix is the one-dimensional subspace $\mathbf{W}_S$ spanned by $\mathbf{e}_2$, while the unstable subspace $\mathbf{W}_U$ is spanned by $\mathbf{e}_1$. Looking at G we see that the trajectory of any point of the form (u, 0) goes to the origin. Hence, the unstable manifold of F is $\mathbf{M}_U = \mathbf{W}_U$.

The stable subspace $\mathbf{W}_S$ of $F_x(\mathbf{0})$ is not the stable manifold of F, because the trajectory of an initial condition $\mathbf{x}_0 = (0, y)$, with $y \neq 0$, goes to infinity.

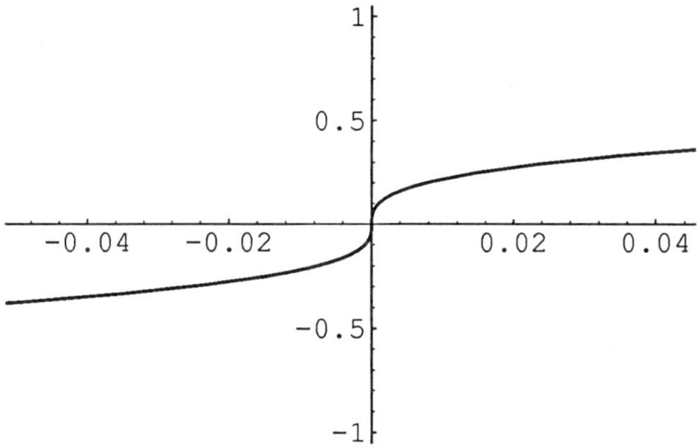

Fig. 5.3.2 The cubic $x = \frac{8}{15}y^3$ is the stable manifold of the dynamical system governed by the function $F(x, y) = (2x - y^3, 0.5\, y)$ at the origin.

To find $\mathbf{M}_S$ let us select an initial condition (x, y) and find what relation between the two coordinates x and y will insure that the orbit O((x, y)) goes to $\mathbf{0}$. After n+1 iterations we obtain

$$(x_{n+1}, y_{n+1}) = (2^n[2x - y^3 - 4^{-2}y^3 - \dots - 4^{-2n}y^3],\ 2^{-n-1}y).$$

The second component of this vector goes to zero, regardless of the value of y. The first one will go to zero if

$$2x = y^3 + 4^{-2}y^3 + \dots + 4^{-2n}y^3 + \dots = y^3[\,1 + 4^{-2} + 4^{-4} + \dots] = \frac{16}{15}y^3 \,.$$

In fact, in this case

$$2x - (y^3 + 4^{-2}y^3 + ... + 4^{-2n}y^3) = \frac{16}{15} y^3 4^{-2(n+1)} ,$$

and the product $2^n 4^{-2(n+1)} \to 0$ as $n \to \infty$. Notice that $\frac{16}{15}$ is the sum of the geometric series $1 + 4^{-2} + 4^{-4} + ...$.

The stable manifold $\mathbf{M_S}$ of F at the origin is the curve $x = \frac{8}{15} y^3$.

As we see, even in a simple case like Example 5.3.5, the problem of determining the stable and unstable manifold at a saddle point is not trivial. It is worth underlining once more the behavior of a saddle equilibrium (see a similar comment in the previous chapter), namely that every initial condition outside the stable manifold gives a trajectory, which in the long run will move away from the equilibrium.

In Example 5.3.5 trajectories starting from every initial condition outside the curve $x = \frac{8}{15} y^3$ will go to infinity. The closer we are to the curve, the longer it takes to detect this behavior. At the beginning we may have the impression that the trajectory is moving towards the origin, but eventually the states will move closer to the unstable manifold and the distance between them and the origin will start increasing.

The behavior of a dynamical system close to a saddle equilibrium is very important in the study of the global behavior of the system. A trajectory leaving a saddle point along its unstable manifold, and coming back to it along its stable manifold, is called **homoclinic** . A trajectory connecting two saddle points is called **heteroclinic**. About hundred years ago H. Poincaré discovered that dynamical systems with heteroclinic of homoclinic trajectories are frequently very complex.

Section 4. Bounded Invariant Sets

In this section we study two classes of dynamical systems, dissipative and quasi-bounded, for which we can determine bounded regions Q in $\mathbf{R}^q$ that are invariant. Any trajectory starting from a point of Q will remain in Q.

Systems of this type are common as mathematical models of real dynamical processes, since "going to infinity" would mean that the system "blows up", a feature not so frequently encountered in real situations.

1.Dissipative maps

The elements of the first class are one-parameter family of maps of the form

(5.4.1) $$F(\mathbf{x}, h, \mathbf{a}) = \mathbf{x} + hG(\mathbf{x}, \mathbf{a})$$

which can be considered one-step Euler's discretizations of the continuous system

(5.4.2) $$\mathbf{x}' = G(\mathbf{x}, \mathbf{a}),$$

with $h = \Delta t$. We assume that G is differentiable with continuous derivative. Given a positive real number r, we can choose h so small that hG is a contraction on the closed ball $D(\mathbf{0}, r)$ centered at the origin and with radius r. In fact the continuity of $G_\mathbf{x}$ implies the existence of a positive number R, such that $\|G_\mathbf{x}(\mathbf{x}, \mathbf{a})\| \leq R$ for all $\mathbf{x} \in D(\mathbf{0}, r)$ (see Lemma 5.1.1). Hence, by selecting h so small that $hR < 1$, we obtain, using the Mean Value Inequality, the desired property for hG.

We say that a map G is of **dissipative** or **point-dissipative** type if there exist two positive numbers r and k such that

(5.4.3) $$G(\mathbf{x}, \mathbf{a}) \bullet \mathbf{x} \leq -k$$

for every vector $\|\mathbf{x}\| \geq r$. The parameter $\mathbf{a} = (a_1, \ldots, a_m)$ is considered fixed. Theorem 5.4.1 below is valid for all values of $\mathbf{a}$ for which (5.4.3) holds.

For dynamical systems governed by maps of the form (5.4.1) and in which the G component is of dissipative type , we have the following result.

Theorem 5.4.1 *Let* F *be of the form* (5.4.1) *and assume that* G *satisfies* (5.4.3) *for some* $r, k > 0$. *Then, given any real number* $R \geq 2r$ *we can find* $d > 0$ *such that for all* $h \leq d$ *the closed closed ball* $D = D(\mathbf{0}, R)$ *centered at the origin and with radius* R *is invariant under the dynamical system defined by* $F(. , h, \mathbf{a})$.

<u>Proof</u>. Choose $R \geq 2r$ and let $L = \max\{\|G(\mathbf{x}, \mathbf{a})\| : \|\mathbf{x}\| \leq R\}$. We first check the size of $F(\mathbf{x}, h, \mathbf{a})$ for all vectors $\mathbf{x}$ such that $r \leq \|\mathbf{x}\| \leq R$. We have

$$\|\mathbf{x} + hG(\mathbf{x}, \mathbf{a})\|^2 = \|\mathbf{x}\|^2 + 2hG(\mathbf{x}, \mathbf{a})\cdot\mathbf{x} + h^2\|G(\mathbf{x}, \mathbf{a})\|^2 \leq \|\mathbf{x}\|^2 + h(- 2k + hL^2)$$

and

$$\|\mathbf{x}\|^2 + h(- 2k + hL^2) \leq R^2 + h(- 2k + hL^2) .$$

Since we want $\|F(\mathbf{x}, h, \mathbf{a})\| \leq R$ we choose d so that $dL^2 < 2k$. This is our first constraint on d.

Now let us choose a vector $\mathbf{x}$ with norm less than r, $\|\mathbf{x}\| < r$. Then

$$\|F(\mathbf{x}, h, \mathbf{a})\| \leq \|\mathbf{x}\| + h\|G(\mathbf{x}, \mathbf{a})\| \leq r + hL.$$

Since we want $\|F(\mathbf{x}, h, \mathbf{a})\| \leq R$ we choose d so that $dL \leq r$. Hence, by selecting d so that both inequalities $dL \leq r$ and $dL^2 \leq 2k$ are satisfied, we obtain the desired result.

Example 5.4.1 The Lorenz model for the study of weather patterns is

$$\begin{aligned} x' &= - ax+ ay \\ y' &= - xz + rx-y \\ z' &= xy - bz . \end{aligned}$$

Use the substitution $u = z - a - r$. The qualitative behavior of the system is not affected by this change since $u' = z'$. We obtain

$$\begin{aligned} x' &= - ax + ay \\ y' &= - xu - ax - y \\ u' &= xy - bu - ab - br . \end{aligned}$$

Euler's one-step discretization gives

$$\begin{aligned} x_{n+1} &= x_n + ah(y_n - x_n) \\ y_{n+1} &= y_n + h(-x_n u_n - a x_n - y_n) \\ u_{n+1} &= u_n + h(x_n y_n - b u_n - b(a + r)) . \end{aligned}$$

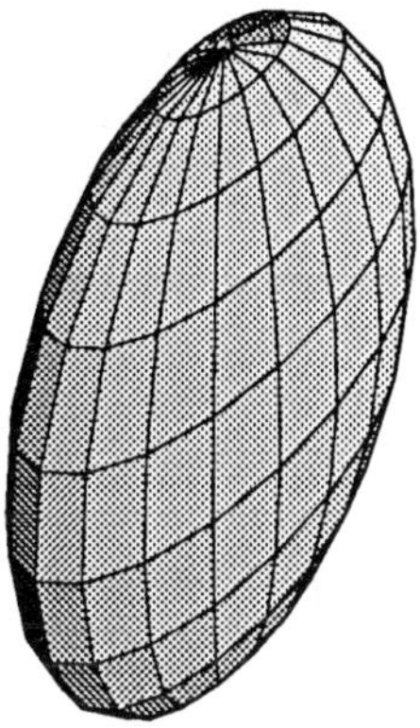

Fig. 5.4.1 The Ellipsoid of the Lorenz map for $a = 8$, $b = 1$ and $r = 16$ in the space xyz. Its sections with planes perpendicular to the x-axis ($x = k$) are circles centered at the point $(k, 0, 12)$. Its sections with planes perpendicular to the y-axis, $y = k$, are ellipses centered at $(0, k, 12)$ with major axis along z and minor along x. The ellipsoid contains the origin and the point $(0, 0, 24)$.

We obtain ($\mathbf{x} = (x, y, u)$, $\mathbf{a} = (a, b, r)$)

$$G(\mathbf{x}, \mathbf{a}) = G((x, y, u), (a, b, r)) = (a(y - x), -xu - ax - y, xy - bu - b(a + r)) ,$$

and

$$G(\mathbf{x}, \mathbf{a})\bullet\mathbf{x} = axy - ax^2 - xuy - axy - y^2 + xyu - bu^2 - b(a + r)u = -ax^2 - y^2 - bu^2 - b(a + r)u .$$

This can be rewritten in the following way

$$G(\mathbf{x}, \mathbf{a})\bullet\mathbf{x} = -(ax^2 + y^2 + b(u + \frac{a + r}{2})^2) + b\frac{(a + r)^2}{4} .$$

Set

$$q^2 = b\frac{(a+r)^2}{4}$$

and consider the ellipsoid EL : $G(\mathbf{x}, \mathbf{a})\bullet\mathbf{x} = 0$, i.e.

$$ax^2 + y^2 + b(u + \frac{a+r}{2})^2 = q^2 .$$

The ellipsoid EL separates the remaining space into two open regions, the interior IN of EL, and the exterior EX of EL. For every vector $\mathbf{x} \in$ IN we have $G(\mathbf{x}, \mathbf{a})\bullet\mathbf{x} > 0$, while for every $\mathbf{x} \in$ EX we have $G(\mathbf{x}, \mathbf{a})\bullet\mathbf{x} < 0$, and the quantity becomes more and more negative as we move away from the ellipsoid. Hence the condition of Theorem 5.4.1 is satisfied by the system obtained via the one-step Euler's discretization of the Lorenz model.

2. **Quasi-bounded maps**

We now give another condition which insures that a discrete dynamical system has a bounded invariant set. First we need to introduce the following idea.

Let $H : \mathbf{R}^q \to \mathbf{R}^q$ and $r > 0$. Denote by $D^c(r)$ the set of vectors of norm not smaller than r, $D^c(r) = \{\mathbf{x} \in \mathbf{R}^q : \|\mathbf{x}\| \geq r\}$. Define

$$N(r) = \sup \{\|H(\mathbf{x})\|/\|\mathbf{x}\| \ : \mathbf{x} \in D^c(r)\}.$$

Notice that $r_1 < r_2$ implies $D^c(r_2) \subset D^c(r_1)$ and $N(r_2) \leq N(r_1)$.

Hence $N(r)$ is not increasing as $r \to \infty$. We may have $N(r) = \infty$ for all $r > 0$. In the case when $N(r) < \infty$, we say that H is **quasi-bounded.**

Example 5.4.2 Let $H_1(x) = x^2$, and $H_2(x) = \ln(1+x^2)$.
The map H_1 is not quasi-bounded . In fact $N(r) = \infty$ for every $r > 0$.
For the map H_2 we have $N(r) \to 0$ as $r \to \infty$. Thus H_2 is quasi-bounded.

In the follwing theorem the parameter **a** is considered fixed. The result is valid for all values of **a** for which the required condition is satisfied.

Theorem 5.4.2 *Let* $F(\mathbf{x}, \mathbf{a})$, $\mathbf{x} \in \mathbf{R}^q$, $\mathbf{a} \in \mathbf{R}^m$, *be quasi-bounded with respect to* $\mathbf{x}$ *and assume that there exists* $r_0 > 0$ *such that* $N(r_0) \leq 1$. *Then there exists a real number* d, *such that for all* $R \geq d$ *the closed ball* $D(R) = \{\mathbf{x} : \|\mathbf{x}\| \leq R\}$ *is invariant under* F.

Proof. Since $N(r)$ is not increasing we have $N(r) \leq 1$ for all $r \geq r_0$. Moreover the continuity of the function $g(\mathbf{x}, \mathbf{a}) = \|F(\mathbf{x}, \mathbf{a})\|$ implies the existence of a constant K such that $\|F(\mathbf{x}, \mathbf{a})\| \leq K$ for all $\|\mathbf{x}\| \leq r_0$.

Choose d equal to the largest of the two numbers r_0 and K. Then, for every real number $R \geq d$ we have

- $\|F(\mathbf{x}, \mathbf{a})\| \leq K \leq R$ for every $\|\mathbf{x}\| < r_0$;
- $\|F(\mathbf{x}, \mathbf{a})\| \leq \|\mathbf{x}\| \leq R$ for every $r_0 \leq \|\mathbf{x}\| \leq R$.

Hence, D(R) is invariant.

Example 5.4.3 Let $F(\mathbf{x}, \mathbf{a}) = F((x,y), (a,b)) = \left(\frac{ay}{1 + x^2}, \frac{bx}{1 + y^2}\right)$.

We find that

$$\frac{\|F(\mathbf{x}, \mathbf{a})\|^2}{\|\mathbf{x}\|^2} = \frac{(ay)^2}{x^2 + y^2}\frac{1}{(1 + x^2)^2} + \frac{(bx)^2}{x^2 + y^2}\frac{1}{(1 + y^2)^2}.$$

If $|a| \leq 1$ and $|b| \leq 1$, the condition of Theorem 5.4.2 is satisfied by the discrete dynamical system governed by F.

Example 5.4.4 As mentioned in the first chapter, the dynamic evolution of a neural net following an initial input $\mathbf{x}_I$, which may remain active during the entire process, can be modelled by the system of equations

$$(5.4.4) \qquad x_{k,n+1} = x_{k,n} - c_k x_{k,n} + t_{k,1} f(x_{1,n}) + \ldots + t_{k,N} f(x_{N,n}) + x_{I,k}$$

for k = 1, 2, ..., N, where f is the neuron response function, and $c_k \in [0, 1]$ for every k. Written in vector form (5.4.4) becomes

$$(5.4.5) \qquad \mathbf{x}_{n+1} = (I - C)\mathbf{x}_n + TG(\mathbf{x}_n) + \mathbf{x}_I,$$

where I is the identity, C is a diagonal matrix, T is called the connectivity matrix and G is a nonlinear map with the property $\|G(\mathbf{x})\| \leq q$ for all $\mathbf{x} \in \mathbf{R}^q$. In Chapter 7 we prove that the dynamical system defined by the map

$$H(\mathbf{x}) = (I - C)\mathbf{x} + TG(\mathbf{x})$$

is quasi-bounded and we give conditions to insure that $N(r_0) \leq 1$ for r_0 sufficiently large.

Section 5. Bifurcation

Dynamical systems in $\mathbf{R}^q$, $q > 1$, offer a larger variety of bifurcation phenomena than one-dimensional systems. They provide, in a more complex setting, the same types of bifurcation points, and they have additional types which are not found in **R**. The following example gives an illustration of bifurcation phenomena similar to the ones occurring in **R**.

Example 5.5.1 Let

$$\begin{cases} x_{n+1} = ax_n(1 - x_n - y_n) \\ y_{n+1} = bx_ny_n . \end{cases}$$

We have $F(\mathbf{x}, \mathbf{a}) = F((x,y), (a,b)) = (ax(1 - x - y), bxy)$ and we see that $(0, 0)$ is always an equilibrium point regardless of the values of the parameters a and b. From

$$\begin{cases} x = ax(1 - x - y) \\ y = bxy \end{cases}$$

we obtain, with $y = 0$, the branch of equilibria

$$\mathbf{x}_s(a) = (x_1(a), y_1(a)) = (\frac{a - 1}{a}, 0) .$$

For $y \neq 0$ we obtain

$$\mathbf{x}_s(a, b) = (x_2(b), y_2(a, b))= (\frac{1}{b}, \frac{ab - a - b}{ab}).$$

It can be easily verified that the $\mathbf{0}$ equilibrium point is a sink for $a \in (0, 1)$, and it is a saddle for $a > 1$. Selecting $b = a$ we obtain that along the branch $(x_1(a), y_1(a))$ the eigenvalues of $F_{\mathbf{x}}$ are $r_1 = 2 - a$, $r_2 = a - 1$. In fact,

$$F_{\mathbf{x}}((x_1(a), x_2(a)), (a, a)) = \begin{pmatrix} 2 - a & 1 - a \\ 0 & a - 1 \end{pmatrix}$$

whose characteristic polynomial is $P(r) = (r - 2 + a)(r - a + 1)$. Thus, the stationary state $(x_1(a), y_1(a))$ is a sink for $a \in (1, 2)$. At this value of a the stability is passed on to the equilibria of the branch $(x_2(a), y_2(a, a))$.

Notice that the two branches intersect each other at $a = 2$, while at $a = 1$ we have $\mathbf{x}_s(a) = \mathbf{0}$. Thus $a = 1$ and $a = 2$ are both transcritical bifurcation points.

Even the simple map of Example 5.5.1 shows that the analysis of bifurcation points in $\mathbf{R}^q$, $q > 1$, is more complex than the corresponding analysis in the one-dimensional case. We shall now mention three typical cases, and illustrate them with some examples. The theoretical justification of the results presented here will be omitted since it goes beyond the purpose of this book. We shall limit our study to one-parameter family of maps. The parameter will be denoted by r.

1. **Bifurcation from the parameter space**

In this first case we assume that the dynamical system has the form

(5.5.1) $$\mathbf{x}_{n+1} = T(\mathbf{x}_n, r) = rM\mathbf{x}_n + F(\mathbf{x}_n, r),$$

where M is a matrix of size q. The function F and its derivative $F_\mathbf{x}$ are continuous. Moreover, we assume that $F(\mathbf{0}, r) = \mathbf{0}$ and $F_\mathbf{x}(\mathbf{0}, r)$ is the zero matrix for every $r \in \mathbf{R}$. Consequently, the origin $\mathbf{x}_s = \mathbf{0}$ is a stationary state for every value of r.

The branch $\mathbf{x}_s(r) = \mathbf{0}$ for all $r \in \mathbf{R}$ is called the *trivial* branch of (5.5.1). We are interested in finding non-zero stationary states. More precisely we want to answer the following question .

Is there any value r_0 such that there exists a continuous function $\phi : J \to \mathbf{R}^q$, where J is an interval, $r_0 \in J$ and the interior of J is not empty, with the following properties

- $\phi(r_0) = \mathbf{0}$;
- $\phi(r) \neq \mathbf{0}$ for all $r \in J$, $r \neq r_0$;
- $\phi(r) = T(\phi(r), r)$ for every $r \in J$?

In this case we say that r_0 is a bifurcation point and the curve

$$\phi : J \to \mathbf{R}^q, \quad r \to \phi(r)$$

is called a bifurcating branch of non-trivial stationary states of $T(. , r)$. Notice that the situation presented in Example 5.5.1 can be reduced to the present setting in the case when $b = a$ $(=r)$ with $M(x, y) = (x, 0)$ and $F((x, y), r) = r(- x^2 - xy, xy)$. Here is another two-dimensional example.

Example 5.5.2 Let

$$T(x, r) = T((x, y), r) = (rx(1+ x), r(- x+2y + y^2)).$$

Then, $M(x, y) = (x, -x + 2y)$ and $F((x, y), r) = r(x^2, y^2)$. We are looking for the non-trivial solutions (x, y) of the system, depending on the parameter r,

$$\begin{cases} x = rx + rx^2 \\ y = -rx + 2ry + ry^2 . \end{cases}$$

For $x \neq 0$, the first equation gives $x = \frac{1-r}{r}$ and from the second we obtain $y = x$, since the solution $y = -1$ cannot be accepted ($\phi(r_0) = \mathbf{0}$).

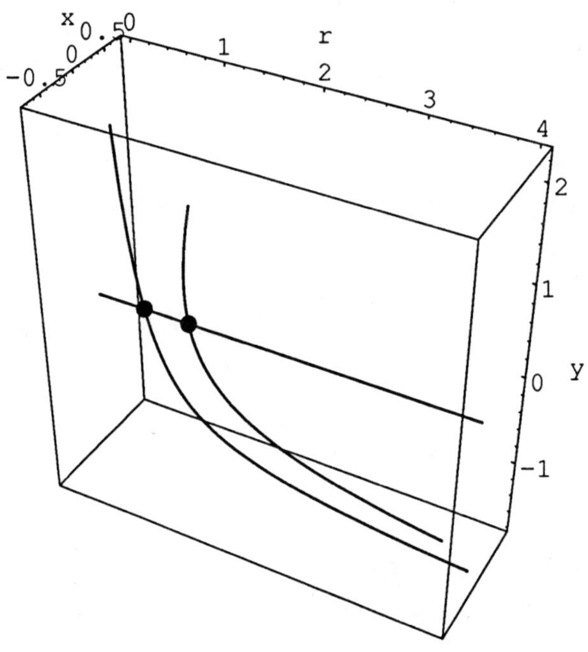

Fig. 5.5.1 Bifurcation diagram of the map $T(x, r) = T((x, y), r) = (rx(1 + x), r(-x + 2y + y^2))$ from the branch of trivial solutions.

The function

$$\phi_1(r) = (r^{-1} - 1, r^{-1} - 1)$$

has the property $\phi_1(1) = (0, 0)$, and we can choose $(0, 2)$ as the open interval J containing the point 1, such that for all $r \in J$ the other two properties are verified. Hence $r_0 = 1$

is a (transcritical) bifurcation point.

For $x = 0$, $y \neq 0$, we obtain from the second equation $y = r^{-1} - 2$. The function

$$\phi_2(r) = (0, r^{-1} - 2)$$

has the property $\phi_2(0.5) = (0, 0)$ and in the open interval $J = (0, 1)$, the function ϕ_2 satisfies all the required properties. Thus $r_0 = 0.5$ is another (transcritical) bifurcation point.

Notice that the two bifurcation points are the reciprocals of the eigenvalues of the matrix M. In fact, it can be shown that this condition is necessary for a parameter value r_0 to be a bifurcation point from the branch of trivial solutions, i.e. $(r_0)^{-1} \in \text{SpecM}$ is necessary for r_0 to be a bifurcation point. Having such a property however does not guarantee that r_0 is in fact a bifurcation point as the following example shows.

Example 5.5.3 Let

$$T(\mathbf{x}, r) = r(x, y) - (y^3, -x^3).$$

In this case $M = I$ whose only eigenvalue is 1. However from

$$(x, y) = r(x, y) - (y^3, -x^3),$$

we get $x \neq 0$ if and only if $y \neq 0$. In this case our search for non-trivial solutions requires that $x^4 + y^4 = 0$, whose only solution is $x = y = 0$. Thus the system does not have any branches of non-trivial stationary states.

The <u>necessary condition stated above becomes sufficient</u> if the algebraic multiplicity of $(r_0)^{-1}$ is odd.

Example 5.5.4 Let $\mathbf{x}_{n+1} = T(\mathbf{x}_n, r) = r(x + z, 2y + 6z, y + z) + (rx^2 - y^2, y^2 - z^2, z^3)$.

Then

$$M = \begin{pmatrix} 1 & 0 & 1 \\ 0 & 2 & 6 \\ 0 & 1 & 1 \end{pmatrix}$$

and $F(\mathbf{x}, r) = (rx^2 - y^2, y^2 - z^2, z^3)$. The eigenvalues of M are 1, -1 and 4. They are all simple. Hence the bifurcation points are 1, -1 and 0.25 . In this case it is quite challenging to determine explicitly as functions of r the three different branches of non-zero stationary states.

2. Bifurcation from the state space

We consider dynamical systems of the form

$$(5.5.2) \qquad \mathbf{x}_{n+1} = \mathbf{x}_n - M\mathbf{x}_n + rF(\mathbf{x}_n) = T(\mathbf{x}_n, r)$$

with $\dim \mathrm{Ker} M \geq 1$. We see that for $r = 0$ every vector $\mathbf{x} \in \mathrm{Ker}M$ is a stationary state of our system. These equilibria will again be called *trivial.* We are interested in the non-trivial ones $(r \neq 0)$. More precisely we would like to find those vectors $\mathbf{x}_s \in \mathrm{Ker}M$ such that there exists a continuous function $\gamma : J \to \mathbf{R}^q$, where J is an interval, $0 \in J$ and the interior of J not empty, with the following properties

- $\gamma(0) = \mathbf{x}_s$;
- $M\gamma(r) = rF(\gamma(r))$, for every $r \in J$.

In this case $r_0 = 0$ is a bifurcation point, and $\mathbf{x}_s$ is a bifurcation state. The curve

$$\gamma : J \to \mathbf{R}, \ r \to \gamma(r)$$

is a bifurcating branch of non-trivial stationary states of (5.5.2).

A result we are going to use in the examples states that a **necessary condition** for $\mathbf{x}_s \in \mathrm{Ker}M$ to be a bifurcation state is that $F(\mathbf{x}_s) \in \mathrm{Im}M$. A <u>sufficient condition</u> is that $F_\mathbf{x}(\mathbf{x}_s)$ is one-to-one on KerM and $F_\mathbf{x}(\mathbf{x}_s)\mathbf{y} \notin \mathrm{Im}M$ for all $\mathbf{y} \in \mathrm{Ker}M$, $\mathbf{y} \neq 0$.

Example 5.5.5 Let

$$T(\mathbf{x}, r) = (0, y) + r(x^2 + y, x - y + 1) = (x, y) - (x, 0) + r(x^2 + y, x - y + 1).$$

The kernel of M is $\mathrm{Ker}M = \{(x, y) : x = 0\}$, and the image is $\mathrm{Im}M = \mathrm{span}\{(1, 0)\}$. We have $F(0, y) \in \mathrm{Im}M$ if and only if $y = 1$. Consequently, the only potential bifurcation state in KerM is $\mathbf{x}_s = (0, 1)$.

Moreover,

$$F_\mathbf{x}(0, 1) = \begin{pmatrix} 0 & 1 \\ 1 & -1 \end{pmatrix} = A, \text{ and } A(0, y) = (y, -y),$$

which implies that the derivative $F_x(0, 1) = A$ is one-to-one on KerM , and $\mathbf{Ay} \notin$ ImM for every $\mathbf{y} \in$ KerM, $\mathbf{y} \neq \mathbf{0}$. The sufficient condition is satisfied.

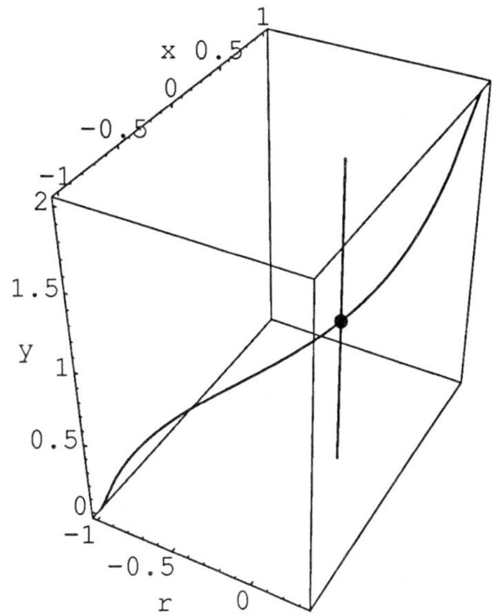

Fig. 5.5.2 The bifurcation diagram of the map $T(\mathbf{x}, r) = (x, y) - (x, 0) + r(x^2 + y, x - y + 1)$, with the non-trivial branch coming out at the point $(0, 1) \in$ KerM.

The bifurcating branch of non-trivial stationary states is the curve $\gamma : (-1, 1/3) \to \mathbf{R}$ defined by

$$\gamma(r) = \begin{cases} \left(\frac{1 - r - \sqrt{1 - 2r - 3r^2}}{2r}, \frac{1 + r - \sqrt{1 - 2r - 3r^2}}{2r}\right) & \text{if } r \neq 0 \\ (0, 1) & \text{if } r = 0 . \end{cases}$$

The function γ is defined in an open interval J containing the origin and it is continuous on J.

The following example shows that the necessary condition stated above may be verified by a state $\mathbf{x}_s$, but bifurcation does not take place.

Example 5.5.6 Let $M : \mathbf{R}^2 \to \mathbf{R}^2$ be the linear map defined by $M\mathbf{x} = M(x, y) = (0, x)$ and $F : \mathbf{R}^2 \to \mathbf{R}^2$ be such that

$$F(x, y) = (x^2 + y^2, x^2 + y^2 - x+1).$$

The kernel of M is the subspace $\mathrm{Ker}M = \{(x, y) : x = 0\}$ and the image is the subspace $\mathrm{Im}M = \mathrm{span}\{(0, 1)\}$. The necessary condition requires that $(y^2, y^2+1) \in \mathrm{Im}M$, or $y = 0$. Consequently $F(0, 0) \in \mathrm{Im}M$ and, obviously, $(0, 0) \in \mathrm{Ker}M$. From

$$(0, x) = r(x^2 + y^2, x^2 + y^2 - x+1) \qquad \text{with } r \neq 0,$$

we derive $x^2 + y^2 = 0$ which implies $x = y = 0$.
The equality $(0, 0) = rF(0, 0) = r(0, 1)$ gives $r = 0$. Hence, the state $\mathbf{x}_s = \mathbf{0}$ satisfies the necessary condition but it is not a starting point of a branch of non-trivial stationary states. Notice that

$$F_{\mathbf{x}}(\mathbf{0}) = \begin{pmatrix} 0 & 0 \\ -1 & 0 \end{pmatrix} = A$$

and $A(0, y) = (0, 0)$ for all y. The derivative $F_{\mathbf{x}}(\mathbf{0})$ is not one-to-one on $\mathrm{Ker}M$.

The analysis of the stability properties of the stationary states along the bifurcating branches can be performed using the eigenvalues of the derivative.

Example 5.5.7 Let $T(\mathbf{x}, r)$ be the one-parameter family of maps of Example 5.5.2, and let us consider only $r \geq 0$. From

$$T_{\mathbf{x}}(\mathbf{0}, r) = \begin{pmatrix} r & 0 \\ -r & 2r \end{pmatrix},$$

whose eigenvalues are r and $2r$, we derive that the $\mathbf{0}$ stationary state is an attractor for $r \in (0, 0.5)$, a saddle for $r \in (0.5, 1)$ and a repeller for $r > 1$. Along the branch given by $\mathbf{x}_s(r) = (0, r^{-1} - 2)$ and setting $r = 0.5 + \varepsilon$, we have

$$T_{\mathbf{x}}(\mathbf{x}_s(r), r) = \begin{pmatrix} 0.5 + \varepsilon & 0 \\ -0.5 - \varepsilon & 1 - 2\varepsilon \end{pmatrix},$$

whose eigenvalues are $0.5 + \varepsilon$ and $1 - 2\varepsilon$. Thus the equilibrium point $\mathbf{x}_s(r)$ is an attractor for $\varepsilon \in (0, 0.5)$, or $r \in (0.5, 1)$.

3. **Hopf bifurcation**

We shall confine our attention to $\mathbf{R}^2$. We consider the one-parameter family of maps in $\mathbf{R}^2$

$$(5.5.3) \qquad \mathbf{x}_{n+1} = F(\mathbf{x}_n, r)$$

with F of class C^2. Denote the eigenvalues of $F_{\mathbf{x}}(\mathbf{x}, r)$ by $\lambda_1(r), \lambda_2(r)$. Assume that there is a branch of equilibrium points $\mathbf{x}_s(r)$ for $r \in (a, b)$, and

$$(5.5.4) \qquad \lambda_j(r) = \alpha(r)[\mathrm{Cos}\beta(r) + (-1)^j \mathbf{i}\ \mathrm{Sin}\beta(r)],\ j = 1,2\ ;$$

$$\beta(r) \in (0, \pi)\ ,\ \alpha(r) > 0 \ \text{ for } r \in (a, b)\ .$$

Since F is of class C^2, the two functions $\alpha(r)$ and $\beta(r)$ are C^1. Let $c \in (a, b)$ be such that $\alpha(r) < 1$ on (a, c) and $\alpha(r) > 1$ on (c, b). For $r = c$ we must have $\alpha(r) = 1$. Assume that

- $\frac{d}{dr}\alpha(r) > 0$ at $r = c$;
- $\frac{\beta(c)}{\pi}$ is irrational or $\frac{\beta(c)}{\pi} = \frac{2m}{n}$, m and n relatively prime and $n \neq 3, 4$.

Then, at $r = c$ we have the so-called Hopf bifurcation. Orbits close to $\mathbf{x}_s(r)$, which is a repeller for $r > c$, are attracted by either an elliptical invariant curve (when $\beta(c)/\pi$ is irrational) or by a periodic orbit (when $\beta(c)/\pi = 2m/n$). In the cases $n = 3$ or $n = 4$ the bifurcation picture is more complicated.

Example 5.5.8 Let $\mathbf{x}_{n+1} = F(\mathbf{x}, r)$ where $F((x, y), r) = (rx(1 - x - y), rxy)$. The values of the parameter r are restricted to the interval $(0, 4]$. There are three branches of stationary states, namely

$$\mathbf{x}_{s1}(r) = \mathbf{0} = (0, 0)\ ,\ \mathbf{x}_{s2}(r) = (1 - r^{-1}, 0) \ \text{ and } \ \mathbf{x}_{s3}(r) = (r^{-1}, 1 - 2r^{-1}).$$

The matrix $F_x(x, r)$ evaluated along the branch $x_{s3}(r)$ gives

$$F_x(x_{s3}, r) = \begin{pmatrix} 0 & -1 \\ r-2 & 1 \end{pmatrix}.$$

The eigenvalues of this matrix are complex for $r > 2.25$ and their modulus is smaller than 1 for $r \in (2.25, 3)$. At $r = 3$ we have a Hopf bifurcation. Moreover, $\beta(3) = \pi/3$ and $n = 6$. The orbits close to $(r^{-1}, 1 - 2r^{-1})$ for $r > 3$ are attracted by a periodic orbit of period 6 which is itself a function of r.

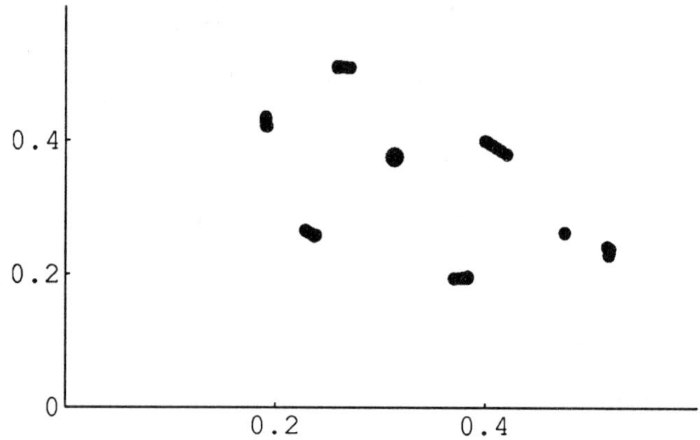

Fig. 5.5.3 The orbit starting at (.4, .1) for $r = 3.2$ converges to a periodic orbit of period 6. The center point is the stationary state $x_{s3}(3.2)$.

This example will be further studied in Chapter 7.

The Hopf bifurcation does not have a counterpart in the one-dimensional case. This topic, as well as the other two presented before, require a mathematical background which is more sophisticated than the one on which this book is based. We have mentioned them here to make our readers aware that discrete dynamical systems have a lot more to say than what we can include in these notes. Upon acquiring greater mathematical skill, the reader will be able to gain further insight into this fascinating field.

Problems

Section 1.

5.1.1 Let $F : \mathbf{R}^q \to \mathbf{R}^q$ be a contraction. Prove that the dynamical system generated by F has at most one equilibrium state.

5.1.2 Prove that the result of Problem 5.1.1 still holds if F is not a contraction, but one of its iterates is.

5.1.3 Let $F : \mathbf{R}^q \to \mathbf{R}^q$ be a contraction with constant $k < 1$ and let $\mathbf{x}_s$ be the only fixed point of F. Consider the trajectory starting from a point $\mathbf{x}_0$. Prove that

$$\|\mathbf{x}_{n+1} - \mathbf{x}_n\| \to 0 \text{ as } n \to \infty .$$

5.1.4 Let $F : \mathbf{R}^q \to \mathbf{R}^q$ be a contraction with constant $k < 1$. Let $\mathbf{x}_0$ be any initial state. Prove the following inequality :

$$\|\mathbf{x}_n - \mathbf{x}_s\| \leq \frac{k^n}{1 - k} \|\mathbf{x}_1 - \mathbf{x}_0\|$$

(Hint : we have $\|\mathbf{x}_0 - \mathbf{x}_s\| \leq \|\mathbf{x}_0 - \mathbf{x}_1\| + \|\mathbf{x}_1 - \mathbf{x}_s\|$ and $\|\mathbf{x}_1 - \mathbf{x}_s\| \leq k\|\mathbf{x}_0 - \mathbf{x}_s\|$) .

This result allows us to estimate how close we are to the equilibrium point without knowing it.

5.1.5 Let $F(x, y) = (0.5x + y^2, 0.3\, y)$. Show that every orbit goes to zero.

5.1.6 Let $F(x, y) = (\mathrm{Cos}\, x + y^2, \mathrm{Cos}\, y)$. Show that the dynamical system governed by the map F has only one stationary state (x_s, y_s). Numerically explore the behavior of several different orbits and verify that they all converge to the unique equilibrium point (x_s, y_s).

5.1.7 Let M be a square matrix of size q. Assume that the spectral radius of M is smaller than 1. Using the equality

$$\rho(M) = \lim_{n \to \infty} \sqrt[n]{\|M^n\|} ,$$

prove that for some integer n, the operator norm of M^n, using the Euclidean norm in $\mathbf{R}^q$, is smaller than 1. Derive from this result that the action of M^n in $\mathbf{R}^q$ is a contraction.

5.1.8 Let M be as in the previous problem. Using Caccioppoli's version of the Banach Contraction Principle, prove that every orbit of the dynamical system governed by M converges to the origin. This provides another way of establishing the result obtained in the previous chapter about the convergence to $\mathbf{0}$ of every orbit of the dynamical system defined by M.

Section 2.

5.2.1 Let $F(\mathbf{x}) = (x^2 - y, -y^3 + 5z, z/2 + x^2)$. Find an equilibrium point of F different from $\mathbf{0}$ and analyze its stability.

5.2.2 Let $F(x, y) = (x^2 + 2y, y^2 + 3x^2)$. Find the eigenvalues of $F_{\mathbf{x}}(0, 0)$ and verify that the point $(0, 0)$ is a sink.

5.2.3 Let $F(\mathbf{x}, a) = (ax + y, ay - xz + 0.01x, xy + 0.9z)$. For which values of the parameter a is the origin an attractor?

5.2.4 Let $F(\mathbf{x}, r) = (0.9x + 0.1y, 0.99\,y + \frac{r}{100}x - 0.01xz, 0.01xy + 0.97334z)$. For which values of $r > 0$ is the origin a sink ?

5.2.5 Let $F(x, y) = (1.4\,x - .2xy, .96y + .2\,xy)$. Find the equilibrium points of this system. Verify that the origin is not a sink.

5.2.6 Let $F(x, y) = (y^2 - 2x + 1, 2x - 3y^2 + 1)$. Verify that the trajectory starting at $(0, 0)$ is periodic of period 2. Is the trajectory an attractor ?

5.2.7 Let $F(x, y) = (x\mathrm{Cos}\,y - y\mathrm{Sin}\,x, x\mathrm{Sin}\,y - y\mathrm{Cos}\,x)$. Verify that the orbit starting at the point (π, π) is periodic . Is the orbit an attractor ?

Section 3.

5.3.1 Let $F(x, y) = (x^2 + 2y, y^2 + 3x)$. Show that $(0, 0)$ is a stationary state. Verify that $(0, 0)$ is a repeller.

5.3.2 Let $F(x, y) = (2x + y^2 - 2, x^2 - x - y^3)$. Show that the point $(1, 1)$ is periodic of period 2. Compute the eigenvalues of F^2 at the point $(1, 1)$. Conclude that the periodic orbit is a repeller.

5.3.3 Let $F(\mathbf{x}, r) = (0.9x + 0.1y, 0.99y + \frac{r}{100}x - 0.01xz, 0.01xy + 0.97334z)$. Show that for $r > 1$ the origin is a saddle.

5.3.4 The Hénon map is defined by $F(\mathbf{x}, \mathbf{a}) = F((x,y), (a,b)) = (1 - ax^2 + y, bx), a, b > 0$. Find the equilibrium points of the Hénon map and check their stability in the case when the parameter $b = 1$.

5.3.5 Consider again the Hénon map with $a = 1$. Check the stability of the equilibrium points.

5.3.6 Verify that the map $F(x, y) = (2x - y^3, \frac{1}{2}y)$ is one-to-one and its inverse is the map $G(u, v) = (\frac{u + 8v^3}{2}, 2v)$.

5.3.7 Verify that after $n+1$ iterations the orbit starting from (x, y), and governed by the map F of the previous problem, reaches the state

$$(x_{n+1}, y_{n+1}) = (2^n[2x - y^3 \text{-} 4^{-2}y^3 \text{-} \ldots - 4^{-2n}y^3],\ 2^{-n-1}y).$$

5.3.8 Show that the trajectory (x_{n+1}, y_{n+1}) of Problem 5.3.7 will go to zero if the initial condition satisfies the equality $x = \frac{8}{15}y^3$, by verifying that

$$2x - y^3(1 + 4^{-2} + \ldots + 4^{-2n}) = \frac{16}{15}y^3 4^{-2(n+1)}.$$

5.3.9 Let $T : \mathbf{R}^q \to \mathbf{R}^q$ be of the form $T(\mathbf{x}) = \mathbf{x} + F(\mathbf{x})$ and let $\mathbf{x}_0$ be any point of $\mathbf{R}^q$. Prove that every eigenvalue λ of $T_\mathbf{x}(\mathbf{x}_0)$ is of the form $\lambda = 1 + \mu$, where μ is an eigenvalue of $F_\mathbf{x}(\mathbf{x}_0)$.

5.3.10 Using the idea of the previous problem verify that the origin is a saddle point of the Lorenz map whenever $r > 1$.

5.3.11 Verify that the origin is a saddle point of the predator-prey model for competition of species without harvesting (see section 2 of Chapter 1). Verify that in the presence of "heavy" fishing the origin may become an attractor.

Section 4.

5.4.1 Find the ellipsoid of the Lorenz map for

$$a = 10,\ b = \frac{8}{3},\ \text{and}\ r = \frac{470}{19}.$$

Check if the equilibrium points belong to the ellipsoid. Verify that for every $\mathbf{x} = (x, y, u)$ inside the ellipsoid, we have $G(\mathbf{x}, \mathbf{a})\bullet\mathbf{x} > 0$, $\mathbf{a} = (10, \frac{8}{3}, \frac{470}{19})$.

5.4.2 Let $F : \mathbf{R}^q \to \mathbf{R}^q$ be a contraction, with constant k. Assume that $F(\mathbf{x}_s) = \mathbf{x}_s$. Verify that every ball centered at $\mathbf{x}_s$ is invariant under F. Find an invariant ball around the origin.

5.4.3 Let $F(\mathbf{x}) = f(\|\mathbf{x}\|)\, G(\mathbf{x})$, where $f : \mathbf{R} \to \mathbf{R}$ is such that

$$f(t) = \ln(1 + t^2),\ \text{and}\ \|G(\mathbf{x})\| \le 0.5\,\|\mathbf{x}\|.$$

Find a ball which is mapped into itself by F.

5.4.4 Verify that the map of Example 5.4.3 satisfies the condition of Theorem 5.4.2, provided that $|a| \le 1$ and $|b| \le 1$.

5.4.5 Verify that the map defined by a matrix M is always quasi-bounded, and it satisfies the condition of Theorem 5.4.2 with respect to a suitable norm, provided that the spectral radius of M is less than 1.

5.4.6 Assume that the matrix M is symmetric and its spectral radius is 1. Using the Euclidean norm, prove that every closed ball D centered at the origin is invariant.

5.4.7 Let $F : \mathbf{R}^q \to \mathbf{R}^q$ be bounded, i.e. there exists a constant $k > 0$ such that

$$\|F(\mathbf{x})\| \le k \text{ for all } \mathbf{x} \in \mathbf{R}^q.$$

Let T be a matrix of size q. Prove that the composite map TF is quasi-bounded (this idea will be used later on the study of Neural Networks).

5.4.8 Let R, L, h, k be as in Theorem 5.4.1 . Verify that $R^2 + h^2L^2 - 2hk \ge 0$.

Section 5.

5.5.1 Work out the details of Example 5.5.2. In particular verify that the function $\phi(r) = (r^{-1} - 1, r^{-1} - 1)$ satisfies the two properties $\phi(r) \ne \mathbf{0}$, $r \in J$, $r \ne 1$ and that for every real number $r \in J$ we have $\phi(r) = T(\phi(r), r)$.

5.5.2 Let $\gamma(r)$ be as in Example 5.5.5. Verify that γ is continuous at $r = 0$.

5.5.3 Let

$$M = \begin{pmatrix} 1 & 0 \\ 0 & 0 \end{pmatrix},$$

and $F(x, y) = (x^2 + y, x - y + 1)$. With γ as in the previous problem verify that $M\gamma(r) = rF(\gamma(r))$.

5.5.4 Find another example of a one-parameter family of maps which satisfies the necessary but not the sufficient condition for bifurcation from the state space (like the one in Example 5.5.6).

5.5.5 Study the stability of the stationary states along the bifurcating branches of Examples 5.5.2 and 5.5.4.

5.5.6 Let $\mathbf{x}_{n+1} = F(\mathbf{x}_n, r)$ where $F((x, y), r) = (rx(1 - x + y), rxy)$, $r \in (0, 4]$. Find the stationary states and analyze the stability of the origin.

CHAPTER 6

CHAOTIC BEHAVIOR

SUMMARY

In the first section of this chapter we introduce the notion of core *of a dynamical system, and of particular subsets of the* core, *called* attractors. *In section 2 we propose a definition of* chaotic systems *and we compare it with definitions given by other authors. In the third section we talk about the* capacity *and the* correlation dimension *of an attractor. In section 4 we present the* Lyapunov exponents.

Section 1. Attractors

We have already encountered the word *attractor*. We used it as another name for sink, in the case of a stationary state or of a periodic orbit.

Example 6.1.1 Let $F(x) = 2x(1 - x)$. The stationary state $x_s = 0.5$ is an attractor since $F_x(0.5) = 0$. Orbits starting at points close to 0.5 converge to 0.5. Actually, in this case it can be shown that every orbit converges to 0.5 except $O(0)$ and $O(1)$.

Example 6.1.2 Consider the dynamical system $F(x, a) = ax(1 - x)$, $a \in [0, 4]$. Recall that for $a > 3$ there is the unique periodic orbit of period 2

$$x_1(a) = \frac{a + 1 + \sqrt{a^2 - 2a - 3}}{2a}, \quad x_2(a) = \frac{a + 1 - \sqrt{a^2 - 2a - 3}}{2a}.$$

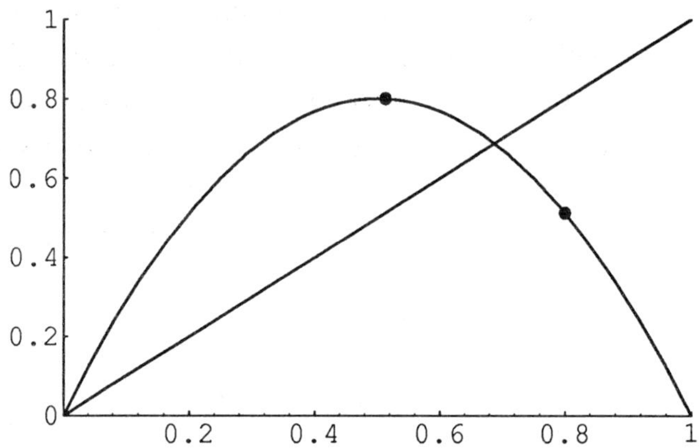

Fig. 6.1.1 The attracting periodic orbit of period 2 of the logistic system for $a = 3.2$. Starting at $x = .3$, 200 iterations were computed and only the last 100 were plotted using the points (x_n, x_{n+1}), $n = 100, \ldots, 199$.

Let $a = 3.2$. Then

$$x_1(3.2) = \frac{2.1 + \sqrt{.21}}{3.2}, \quad x_2(3.2) = \frac{2.1 - \sqrt{.21}}{3.2},$$

and $F_x(x_1(3.2), 3.2)F_x(x_2(3.2), 3.2) = 0.16$. The periodic orbit is an attractor. Orbits with initial state close to $x_1(3.2)$ or $x_2(3.2)$ converge to the periodic orbit (see Fig. 6.1.1)

The analysis of the long range behavior of a dynamical system requires a definition of attractor, which is applicable to more complex sets than stationary states and periodic orbits. This is our purpose in the following pages.

Let $F : D \to D$, where D is a disk or more generally, a bounded region of $\mathbf{R}^q$. Recall that $L(\mathbf{x}_0)$, $\mathbf{x}_0 \in D$, denotes the set of all limit points of the orbit $O(\mathbf{x}_0)$. Assume that $\mathbf{z} \in L(\mathbf{x}_0)$, which means that a subsequence of $O(\mathbf{x}_0)$ converges to $\mathbf{z}$. Then $F(\mathbf{z})$ is the limit of the subsequence of $O(\mathbf{x}_0)$, obtained by adding 1 to the indices of the subsequence converging to $\mathbf{z}$. Consequently, $F(\mathbf{z}) \in L(\mathbf{x}_0)$ and the image of $L(\mathbf{x}_0)$ under F is contained in $L(\mathbf{x}_0)$, i.e. $F(L(\mathbf{x}_0)) \subset L(\mathbf{x}_0)$. Similarly one can show that if $\mathbf{u} \in L(\mathbf{x}_0)$ then $\mathbf{u} \in F(L(\mathbf{x}_0))$, i.e. there exists $\mathbf{x} \in L(\mathbf{x}_0)$ such that $F(\mathbf{x}) = \mathbf{u}$.
Thus $L(\mathbf{x}_0) \subset F(L(\mathbf{x}_0))$. The two inclusions imply $F(L(\mathbf{x}_0)) = L(\mathbf{x}_0)$.

The above equality is true for every $\mathbf{x}_0 \in D$. Therefore the set

$$C = \bigcup_{\mathbf{x} \in D} L(\mathbf{x}) \tag{6.1.1}$$

has the important property $F(C) = C$. We denote by $\mathbf{C}$ the closure of C and we call it the **core** of the dynamical system generated by F.

Example 6.1.3 Let

$$F(x) = \begin{cases} x^2 & \text{if } -3 \le x \le 1 \\ 4\sqrt{x} - 3 & \text{if } 1 < x \le 9. \end{cases}$$

F maps the interval $[-3, 9]$ into itself. The core of F is the set $\mathbf{C} = \{0, 1, 9\}$ (see Fig.6.1.4).

In the evolution of a dynamical system the core $\mathbf{C}$ is the set towards which every orbit moves in the long run. Therefore, the geometry of $\mathbf{C}$ is of primary importance for understanding those characteristics of the system which depend on its long range behavior. Computer generated trajectories can be of significant help in visualizing the core. Cases in which $\mathbf{C}$ is a point or a periodic orbit or the entire set D are usually easier to analyze. There are, however, dynamical systems in which the core, or part of it, assumes forms with very complicated geometry.

Our first step in studying the core is to distinguish its **blocks**. A block A of **C** is a subset of **C** having the following properties:

a) F(A) = A,

b) there exists a positive real number r such that

$$(\bigcup_{x \in A} B(x, r)) \cap \mathbf{C} = A . \quad (6.1.2)$$

In other words A is invariant under the action of F and it is isolated from the rest of **C** in the sense that every point of **C** which is not in A is far from A by at least as much as r.

A block A of **C** is called **a building block** if it does not properly contain any other block of **C**.

Example 6.1.4 The core of the dynamical system of Example 6.1.3 has three building blocks, namely the three points 0, 1, and 9. The set {0, 1} is a block, but not a building block.

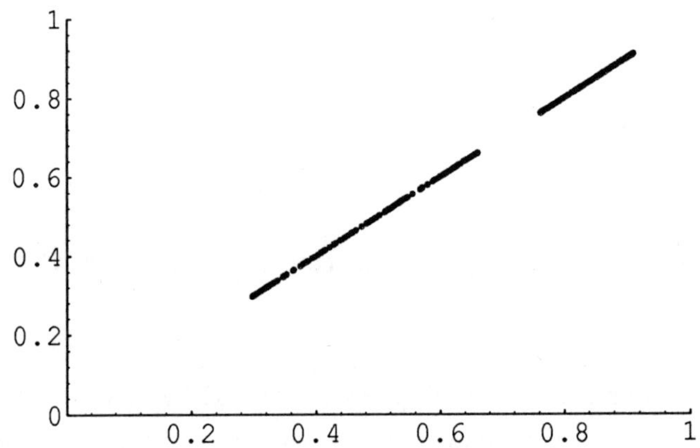

Fig. 6.1.2 The limit set L(.3) of the dynamical system governed by 3.64x(1 - x) is visualized here. We have plotted the sequence of points (x_n, x_n), n = 200, ...,1000. All points visible in the graph appear to be limit points of the sequence.

Definition 6.1.1 A building block **A** of **C** is an **attractor** if there exists $\varepsilon > 0$ such that all trajectories having initial state at distance less than ε from **A** will converge to **A**.

To make this idea more precise define

$$(6.1.3) \qquad d(\mathbf{x}, \mathbf{A}) = \inf\{ \|\mathbf{x} - \mathbf{c}\| : \mathbf{c} \in \mathbf{A}\} .$$

Then **A** is an **attractor** if there exists $\varepsilon > 0$ such that

$$\mathbf{x}_0 \in D \text{ and } d(\mathbf{x}_0, \mathbf{A}) < \varepsilon$$

implies $d(\mathbf{x}_n, \mathbf{A}) \to 0$ as $n \to \infty$.

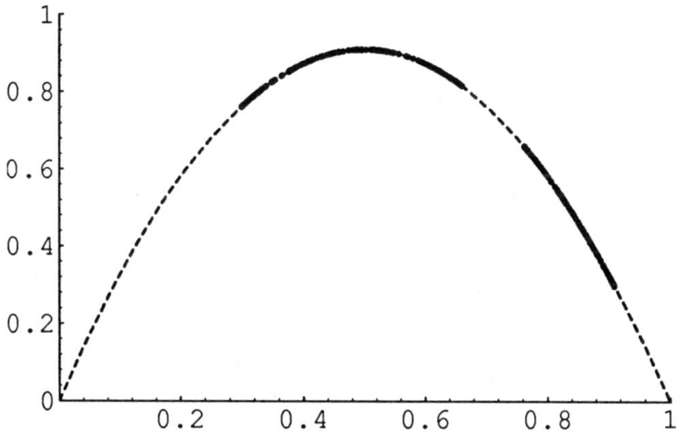

Fig. 6.1.3 To further dramatize the picture of Fig. 6.1.2 we have plotted the sequence (x_n, x_{n+1}). We see that the sequence "reconstructs" a good portion of the parabola $3.64x(1-x)$.

The attracting building blocks of the core are of primary importance in the study of the long range behavior of a system. It is not unusual for their geometry to be complicated to the point that some of them have been called **strange attractors.**

Given an attractor **A**, we call **basin of attraction** of **A** the set of all initial conditions $\mathbf{x}_0$, such that $d(\mathbf{x}_n, \mathbf{A}) \to 0$ as $n \to \infty$.

The various attractors of the core **C** will divide the domain of the dynamical systems into basins of attractions separated by **basin boundaries**. The geometry of these boundaries is frequently at least as complex as the geometry of the attractors themselves.

Example 6.1.5 The stability analysis of the core of Example 6.1.3 shows that the equilibrium point $x_{s1} = 0$ is an attractor and its basin of attraction is the interval $(-1, 1)$. Another attractor is $x_{s2} = 9$ and its basin of attraction are the two intervals $[-3, -1)$ and $(1, 9]$. The fixed point $x_{s3} = 1$ is a repeller (see Fig. 6.1.4).

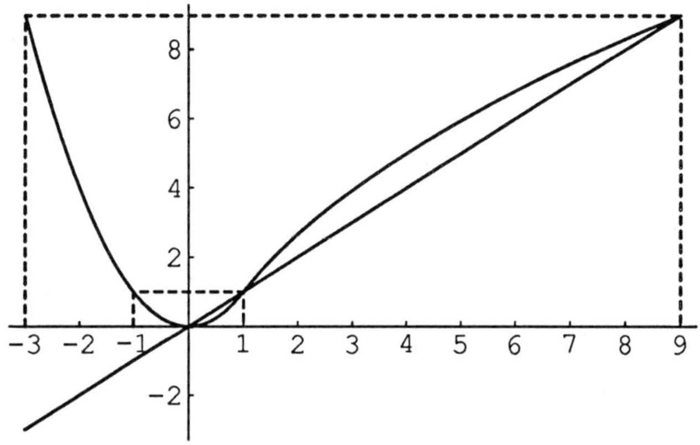

Fig. 6.1.4 This graph shows that the domain of attraction of x_{s2} is the union of two intervals: $[-3, -1)$ and $(1, 9]$, while the domain of attraction of the origin is the interval $(-1, 1)$.

Our definition and study of chaotic behavior focus on the attractors of dynamical systems, leaving aside basins of attraction and their boundaries. In section 2 we propose that **two conditions** be met by an attractor **A** of a dynamical system F to consider the system chaotic on **A**. They can be roughly described as <u>instability</u> of F on **A** and <u>indecomposability</u> of **A**.

Remark 6.1.1 Although the two conditions could be viewed as "minimal", it is nevertheless difficult in many cases to establish theoretically that they are met. Therefore, in sections 3 and 4, we propose to seek the support of numerical evidence based on the correlation dimension and the Lyapunov exponents. Although both ideas are well-suited for testing the two conditions, some preference is given to the correlation dimension as a test of indecomposability and to the Lyapunov exponents as a test of instability.

Section 2. Chaotic Dynamical Systems

1. Stable and unstable orbits

In the first chapter we introduced the idea of **aperiodic** orbit, by requiring that the limit set of the orbit be **infinite.** We want to show here that every orbit of a counter-clockwise rotation $P(\theta)$ of $\mathbf{R}^2$ around the origin, by an angle θ such that θ/π is irrational, and starting from an initial state $\mathbf{x} \neq \mathbf{0}$, has infinitely many limit points.

We assume that the initial state $\mathbf{x}$ is on the unit circle, $\|\mathbf{x}\| = 1$, and prove that every arc, however small, of this circle contains points of the orbit $O(\mathbf{x})$.

In fact, the existence of an arc A with positive length which does not intersect $O(\mathbf{x})$, implies the existence of the largest possible open arc, A_0, containing A such that A_0 does not contain any points of $O(\mathbf{x})$, but its end points belong to $O(\mathbf{x})$. Clockwise rotation of A_0 by θ provides another open arc, A_1, with the same property as A_0, namely no points of $O(\mathbf{x})$ are inside A_1, but the end points of A_1 belong to $O(\mathbf{x})$. Notice that A_0 and A_1 are disjoint. In fact, if $A_0 = A_1$ then $\theta = 2\pi$, which is impossible and if $A_0 \cap A_1 \neq \varnothing$ then A_0 contains one of the end points of A_1, which is impossible as well because both end points of A_1 belong to $O(\mathbf{x})$.

Continuing in this manner we obtain successive arcs $A_2, A_3, \ldots, A_n$, such that none of them can coincide with a previous one, otherwise θ/π would be rational, and all arcs are pair-wise disjoint, otherwise we could find a point of $O(\mathbf{x})$ inside one of them. Eventually there will be so many arcs that their total length will exceed 2π, a result which is obviously absurd.

Thus, every point $\mathbf{z}$ of the circle is a limit point of the orbit $O(\mathbf{x})$, since every open arc, however small, centered at $\mathbf{z}$ contains states of the orbit.

The dynamical system generated by the rotation $P(\theta)$ is linear. Its explicit form is

$$(6.2.1) \qquad \begin{cases} x_{n+1} = x_n \operatorname{Cos}\theta - y_n \operatorname{Sin}\theta \\ y_{n+1} = x_n \operatorname{Sin}\theta + y_n \operatorname{Cos}\theta . \end{cases}$$

We feel that, on intuitive grounds, the system defined by (6.2.1) cannot be considered chaotic, even though all non stationary orbits of the system are aperiodic. In particular we see that orbits whose initial states are close will remain close forever and the evolution of the system from any initial condition $\mathbf{x}_0$ is determined quite accurately even if the state $\mathbf{x}_0$ is not known exactly. Any error in the measurement of the initial coordinates is not magnified by successive iterations. Every orbit of the system is stable.

The conditions which qualify a system as chaotic (see DCS below) require the presence of aperiodic orbits, but are not met by the rotation $P(\theta)$ or by any dynamical system governed by a $q \times q$ matrix M. The presence of nonlinear terms in the equations governing the evolution of the system is necessary to generate chaotic behavior .

Our first step in the direction of defining chaotic systems is to provide the formal definition of stability of not necessarily stationary or periodic orbits, which was mentioned in Chapter 1.

Definition of stable orbit (DSO)

An orbit $O(\mathbf{x}_0)$ of a dynamical system $F : \mathbf{R}^q \to \mathbf{R}^q$ is said to be **stable** if for every $r > 0$ there exists $d > 0$ such that

$$\|\mathbf{y}_0 - \mathbf{x}_0\| \leq d \quad \text{implies} \quad \|\mathbf{y}_n - \mathbf{x}_n\| \leq r \quad \text{for all} \quad n \geq 1.$$

An orbit which is not stable will be called **unstable.**

In other words, $O(\mathbf{x}_0)$ is unstable if there exists $r(\mathbf{x}_0) > 0$ such that for every positive number d we can find an initial state $\mathbf{y}_0$, $\|\mathbf{y}_0 - \mathbf{x}_0\| \leq d$, whose orbit is not contained in the closed ball $D(\mathbf{x}_0, r(\mathbf{x}_0))$.

Notice that the definition of stability given above coincides with the ones given previously for the stability of a stationary state in dimension 1 or higher. In the case of a periodic orbit of prime period p we defined its stability in terms of F^p, by requiring the states of the orbit to be stable stationary states for the dynamical system generated by F^p. We need to reconcile this definition with the one given above.

For simplicity, assume that $p = 2$, and $\mathbf{x}_0, \mathbf{x}_1$ are stable equilibria for F^2, such that $\mathbf{x}_1 = F(\mathbf{x}_0)$, $\mathbf{x}_0 = F(\mathbf{x}_1)$. Let $r > 0$ be fixed. There exist d_0, d_1, with the property

$$\|\mathbf{y}_0 - \mathbf{x}_0\| \le d_0 \ \text{ implies } \ \|\mathbf{y}_{2n} - \mathbf{x}_0\| \le r, \ \text{ and}$$

$$\|\mathbf{z}_1 - \mathbf{x}_1\| \le d_1 \ \text{ implies } \ \|\mathbf{z}_{2n+1} - \mathbf{x}_1\| \le r \ \text{ for all } \ n \ge 1.$$

Since F is continuous we can choose $\rho \le \min\{d_0, d_1\}$ such that

$$\|\mathbf{y} - \mathbf{x}_0\| \le \rho \ \text{ implies } \ \|F(\mathbf{y}) - \mathbf{x}_1\| \le d_1, \ \text{ and } \ \|\mathbf{z} - \mathbf{x}_1\| \le \rho \ \text{ implies } \ \|F(\mathbf{z}) - \mathbf{x}_0\| \le d_0.$$

Select any initial state $\mathbf{y}_0 \in D(\mathbf{x}_0, \rho)$. Then $\mathbf{y}_1 = F(\mathbf{y}_0) \in D(\mathbf{x}_1, d_1)$. Hence $\mathbf{y}_{2n} \in D(\mathbf{x}_0, r)$ and $\mathbf{y}_{2n+1} \in D(\mathbf{x}_1, r)$ for every $n \ge 1$.
Similarly,

$$\mathbf{z}_1 \in D(\mathbf{x}_1, \rho) \ \text{ implies } \ \mathbf{z}_2 = F(\mathbf{z}_1) \in D(\mathbf{x}_0, d_0) \ \text{ and } \ \mathbf{z}_{2n} \in D(\mathbf{x}_0, r), \ \mathbf{z}_{2n+1} \in D(\mathbf{x}_1, r)$$

for every $n \ge 1$.

Therefore, the periodic orbit is stable in the sense of (DSO). Vice versa, it can be shown, using similar arguments, that all points of a periodic orbit of period p which is a stable periodic orbit in the sense of (DSO), are stable stationary states of F^p. Thus, the two definitions are reconciled.

Any orbit of the dynamical system (6.2.1) is stable in the sense of (DSO), since the rotation, being a rigid motion, preserves the distances. We can choose $r = d$. Examples of unstable non stationary and non periodic orbits will be provided below.

2. Chaotic systems

Before presenting our definition of chaotic systems we would like to discuss some recent developments which have motivated the attention reserved today to chaotic dynamics.

In 1963 E.N. Lorenz [Lorenz, 1963] proposed the system of differential equations presented in Chapter 1 as a crude model of atmospheric behavior. In the previous chapter we proved that the discrete dynamical system derived from it with the one-step Euler discretization, has a bounded invariant region in $\mathbf{R}^3$ which attracts all orbits. The same is true for the orbits of the continuous model.

By selecting reasonable values of the control parameters a, b, and r and solving the equations numerically, Lorenz noticed that very minor changes in the initial conditions could produce significantly different outcomes. More precisely, the distance between corresponding states of trajectories very close at $t = 0$, could oscillate aperiodically in an open interval $(0, d)$. Lorenz published the paper "Deterministic nonperiodic flow" to make the scientific community aware of this unexpected behavior. It was suggested that this discovery may imply that no long range prediction of atmospheric changes will ever be possible.

In 1975 two mathematicians T. Li and J.A. Yorke [Li-Yorke, 1975] published a paper in which they proved a surprising theorem. The presence of a periodic orbit of prime period 3 in a dynamical system defined by a continuous map F of an interval I into itself, implies the presence of uncountably many aperiodic and unstable orbits. The title of their paper was "Period three implies chaos."

In 1976, just one year after the discovery of Li and Yorke, the biologist R.M. May [May, 1976] wrote the paper "Simple mathematical models with very complicated dynamics" in which he illustrated the surprisingly complex orbits of a one-dimensional dynamical system in the real line governed by a quadratic map.

Around this time many scientists became interested in dynamical systems having characteristics similar to the one proposed by Lorenz and to the ones analyzed by Li-Yorke and May. The theory of chaotic systems emerged. As R. Devaney [Devaney, 1987] points out, "the field of Dynamical Systems and especially the study of chaotic systems has been hailed as one of the important breakthroughs in science in this century."

Three observations need to be made here. The first is that the use of computer-aided investigation was fundamental in the discovery of Lorenz and has been important in subsequent developments. On one hand, all major ideas and theorems would have certainly been detected and established without the aid of numerical evidence. On the other, the use of computers has certainly played a significant role in providing the theory with supporting evidence.

The second is that H. Poincaré should be credited as the first to propose, about a hundred years ago, a global rather than a local analysis of dynamical systems, and to realize that the presence of homoclinic or heteroclinic trajectories was coupled with very complex behavior. His footsteps were followed in the USA by G.D. Birkhoff [Birkhoff, 1927] with his classical book "Dynamical Systems", and in the Soviet Union by Lyapunov, Pontryagin and others. While the attention of these giants was mainly concentrated on continuous systems, the French mathematicians G. Julia [Julia, 1918] and P. Fatou [Fatou, 1926] were studying the extremely complicated dynamics provided by the iteration of rational functions in the complex plane.

Around 1960, several first-rate scientists gave renewed impulse to the study of Dynamical Systems. Among them we find Smale, Moser, Peixoto, Kolmogorov, Arnold and Sinai. Therefore the field was prepared and ready for the major developments that have taken place in the last thirty years.

The third and final observation is that Chaotic Dynamical Systems have always been an interdisciplinary field. Scientists from areas other than Mathematics have played and continue to play a significant role in the development of the theory. Hardly any mathematical topic has sparked so much cross-cultural interest as this one and the trend continues unabated. Side by side with technical studies we find less formal presentations (see, for example [Stewart, 1989]) and books written for a more general audience (see [Gleick, 1987]).

After these introductory remarks, we propose the definition of discrete chaotic dynamical systems we consider more appropriate to the background required for this book. We also discuss some alternative definitions proposed by other authors. The reader should keep in mind that if we could assume a broader mathematical knowledge, including in particular Measure Theory, then our definition of chaotic systems would be different.

In what follows we assume that $U \subset \mathbf{R}^q$ is open and bounded, $F : \bar{U} \to \bar{U}$ is continuous on $\bar{U}$ and differentiable on U except possibly at finitely many points. We denote by $\mathbf{A}$ an attractor of F in $\bar{U}$.

Definition of Chaotic System (DCS)

The dynamical system defined by F on $\bar{U}$ (or simply F) is chaotic on **A** if

1. every orbit of F in **A** is **unstable**;

2. there exists $\mathbf{x}_0 \in \mathbf{A}$ such that $O(\mathbf{x}_0)$ is **aperiodic** and for every $\mathbf{x} \in \mathbf{A}$ and $r > 0$, we have $B(\mathbf{x}, r) \cap O(\mathbf{x}_0) \neq \varnothing$ (i.e. $O(\mathbf{x}_0)$ is **dense** in **A**).

The above definition looks simple, but it may be difficult to verify that a given system is chaotic in the stated sense, since we need to determine the attractor **A**, to verify the instability of all orbits and to establish the existence of an aperiodic and dense orbit in **A**. Topological conjugacy can be used to make all these steps easier, since it preserves chaotic behavior.

More precisely, let U, V be two open, bounded subsets of $\mathbf{R}^q$, and let F, G

$$F : \bar{U} \to \bar{U}, G : \bar{V} \to \bar{V}$$

be conjugate by a map $h : \bar{U} \to \bar{V}$ (h is continuous, invertible, with continuous inverse). Assume that $\mathbf{A} \subset \bar{U}$ is an attractor of F and that F is chaotic in **A**. Then h(**A**) is an attractor for G, and G is chaotic in h(**A**). In fact, from the equalities

$$G^n(\mathbf{y}) = hF^n(\mathbf{x}), \qquad \mathbf{y} = h(\mathbf{x}),$$

and from the continuity of h, h^{-1} we derive that every orbit of G in h(**A**) is unstable. Moreover, if $O(\mathbf{x}_0)$, $\mathbf{x}_0 \in \mathbf{A}$, is aperiodic and dense in **A**, then $O(\mathbf{y}_0)$, $\mathbf{y}_0 = h(\mathbf{x}_0)$ is aperiodic and dense in h(**A**).

Example 6.2.1 Let $I = [-1, 1]$ and $G : I \to I$ be defined by $G(x) = -2|x| + 1$. No orbit of the system generated by G is stable. In fact the function G is differentiable except at the point 0 and $G(0) = 1$, $G(1) = -1$. Therefore an orbit containing 0 is eventually stationary and its limit point is an unstable fixed point since $G_x(-1) = 2$. At any other point we have $|G_x(x)| = 2$. Hence, no orbit is stable.

Later we shall prove that the core of the system is the entire interval [-1, 1] and that G has a dense (and obviously aperiodic) orbit in [-1, 1]. Thus the dynamical system $x_{n+1} = G(x_n)$ is chaotic on [-1, 1] in the sense of DCS.

Notice that G has a periodic orbit of period 3. In fact

$$G(-5/9) = -1/9, \quad G(-1/9) = 7/9 \quad \text{and} \quad G(7/9) = -5/9.$$

Since G is continuous we can apply the result of Sarkovskii to derive that G has periodic orbits of every prime period.

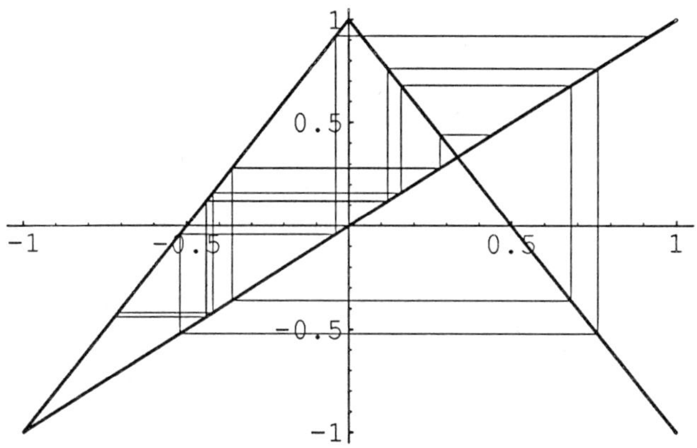

Fig. 6.2.1 The graph of the function G of Example 6.2.1. The instability is graphically illustrated with the cobweb method.

Consider now the system governed by $H : [0, 1] \to [0, 1]$, defined by

$$H(x) = \begin{cases} 2x & \text{if } 0 \leq x \leq \frac{1}{2} \\ 2 - 2x & \text{if } \frac{1}{2} < x \leq 1 . \end{cases}$$

The maps H and G are conjugate via the function $h_1 : [-1, 1] \to [0, 1]$, $h_1(x) = 0.5\,(x+1)$. Consequently, no orbit of H is stable. The core **C** of H is the h_1 image of the core of G. Hence **C** = [0, 1] and H is chaotic on [0, 1]. The map H has periodic orbits of every prime period.

In Chapter 2 we graphically verified that G and the map

$$F : [-1, 1] \to [-1, 1], \quad F(x) = -2x^2+1$$

are conjugate via the function

$$h(x) = \frac{2}{\pi} \text{ArcSin}x .$$

This fact can be established theoretically using the following result from Calculus.

Theorem 6.2.1. *Let* f, g : [a, b] → **R** *be continuous on* [a, b] . *Assume that* f *and* g *are differentiable on* (a, b) *except possibly at finitely many points. Then* $f_x(x) = g_x(x)$ *for every* x *where both derivatives exist, implies the existence of a constant* r *such that* f(x) = g(x) + r *for all* x ∈ [a, b].

With Theorem 6.2.1 we can prove that h(F(x)) = G(h(x)) for every x ∈ [-1, 1] . In fact both the right and left hand side are continuous on [-1, 1] and differentiable on (-1, 1) except at finitely many points. Moreover $h_x(F(x))F_x(x) = G_x(h(x))h_x(x)$.

From h(F(-1)) = G(h(-1)) we obtain r = 0. Consequently, F and G are conjugate, and F is chaotic on [-1, 1]. The function K : [0,1]→ [0,1] , K(x) = 4x(1- x) is conjugate to F with the map $h_1 : [-1, 1] \to [0, 1]$, $h_1(x) = 0.5\,(x+1)$. Thus K is chaotic on [0, 1].

Many authors replace the instability condition included in DCS with a stronger property.

Definition of sensitive dependence on initial conditions (DSD)

The dynamical system defined by F on $\bar{U}$ (or simply F) has in **A** **sensitive dependence** with respect to initial conditions if there exists a positive real number ρ such that no matter what initial condition $\mathbf{x}_0$ we select in **A** and no matter how small is the ball $B(\mathbf{x}_0, r)$, we can find a state $\mathbf{y}_0 \in B(\mathbf{x}_0, r) \cap \mathbf{A}$, such that $\|F^m(\mathbf{y}_0) - F^m(\mathbf{x}_0)\| > \rho$ for some integer $m \geq 1$.

The conditions of instability or sensitivity are frequently verified on the entire basin of attraction Q of **A**, although this constraint is not explicitly imposed. Notice the difference between the two properties. Instability requires that for every $\mathbf{x}_0 \in \mathbf{A}$ there exists a separation constant $r(\mathbf{x}_0)$ such that as close as we like to $\mathbf{x}_0$ we can find an initial condition $\mathbf{y}_0 \in \mathbf{A}$ with the property $\|F^m(\mathbf{x}_0) - F^m(\mathbf{y}_0)\| > r(\mathbf{x}_0)$ for some $m \geq 1$. Sensitivity requires, in addition, $r(\mathbf{x}_0)$ to be the same for all $\mathbf{x}_0 \in \mathbf{A}$.

As we shall see, instability and sensitive dependence on initial conditions of a dynamical system F on an attractor **A** can be tested numerically (see Section 4 on Lyapunov exponents). This analysis cannot be regarded as a fool-proof technique, but it provides some clue about the long term behavior of the system.

A dynamical system F which satisfies DSD on an attractor **A** and has an aperiodic and dense orbit in **A** is chaotic (on **A**) according to DCS, since sensitivity with respect to initial conditions implies instability.
We have opted for the weaker condition required by DCS for the following reasons. On one hand the presence of the common separating factor ρ makes the verification of sensitivity harder than the verification of instability. On the other, the added information does not appear to be of crucial importance in the study of the long range behavior of F.

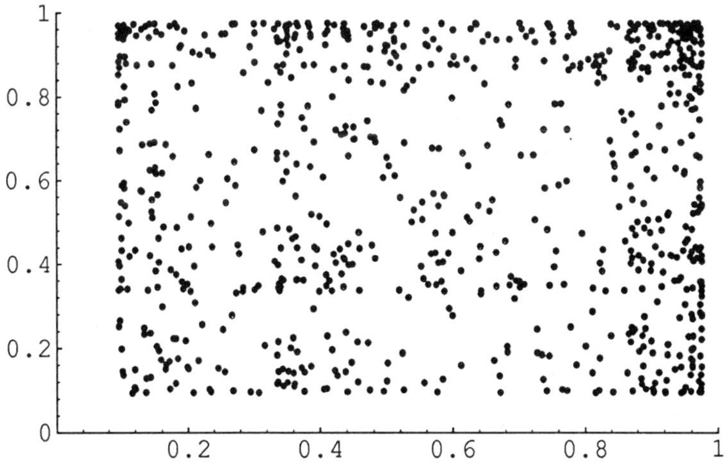

Fig. 6.2.2 The sensitivity with respect to initial conditions of the dynamical system governed by the map $F(x) = 3.9x(1 - x)$ is dramatically underlined by the above graph, in which corresponding states of the trajectories starting from $x_0 = .3$ and $y_0 = .31$ have been plotted versus each other. We see a lot of states close to the diagonal $y = x$ (the corresponding states are close) and a lot of corresponding states as far away as .9 from each other.

The following is an additional example of a chaotic system and has a significant theoretical value. The function governing the system has a discontinuity.

Example 6.2.2 Let $I = [0,1]$ and define $B : I \to I$ by $B(x) = 2x - [2x]$, where $[2x]$ denotes the largest integer less than or equal to $2x$.

Following the pattern of Example 6.2.1 we find that $B_x(x) = 2$ whenever the derivative is defined. Hence, no orbit of this system is stable. We shall prove later that $\mathbf{C} = [0,1]$, and there is a dense orbit in $\mathbf{C}$. Hence F is chaotic in the sense of DCS. The map is called baker's transformation on the unit interval [0,1].

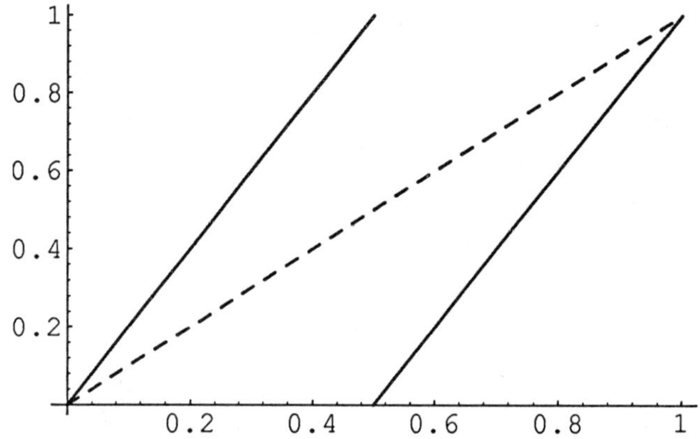

Fig. 6.2.3 The baker's transformation on the unit interval provides one of the most quoted examples of chaotic behavior.

We have mentioned (see Chapter 1) a result obtained by Li and Yorke [Li-Yorke, 1975]. The full version of their theorem states that a dynamical system governed by a continuous function $F : J \to J$, where J is an interval, has uncountably many aperiodic and unstable trajectories, provided that F has a periodic orbit of prime period 3.

In the literature, dynamical systems $F : J \to J$ where F is continuous and with a periodic orbit of prime period 3n for some $n \geq 1$, have been called **chaotic in the Li-Yorke sense.** The idea of Li-Yorke cannot be applied to the map B of Example 6.2.2 .

An extension of the result of Li-Yorke to higher dimension was obtained in 1978 by F.R. Marotto [Marotto, 1978]. He proved that a dynamical system $F : D \to D$, where D is a closed ball in $\mathbf{R}^q$ centered at a stationary state $\mathbf{x}_0$, has uncountably many aperiodic and unstable trajectories if the spectral radius of $F_{\mathbf{x}}(\mathbf{x}_0)$ is larger than 1, and there exists $\mathbf{z}_1$ in the unstable manifold M_U of $\mathbf{x}_0$ such that

$$\det\big((F^m_{\mathbf{x}})(\mathbf{z}_1)\big) \neq 0 \quad \text{and} \quad F^m(\mathbf{z}_1) = \mathbf{x}_0$$

for some positive integer m.

In addition, Marotto proved that there exists an infinite subset S of the natural numbers **N**, such that for every $p \in S$, the dynamical system has a periodic orbit of prime period p.
Marotto called $\mathbf{x}_0$ a **snap-back repeller** .

It is legitimate to ask if the presence of a periodic orbit of period 3 in a one-dimensional dynamical system, or the presence of a snap-back repeller in a dynamical system of dimension higher than 1, both governed by continuous functions, authorizes us to say that the systems are chaotic in the sense of DCS. The following example illustrates the difficulty of answering this question.

Example 6.2.3. Let $F : [0, 1] \to [0, 1]$ be defined by

$$F(x) = \begin{cases} 0.5x & \text{if} \quad 0 \leq x \leq 0.2 \\ 3x - 0.5 & \text{if} \quad 0.2 < x \leq 0.5 \\ 2 - 2x & \text{if} \quad 0.5 < x \leq 1 . \end{cases}$$

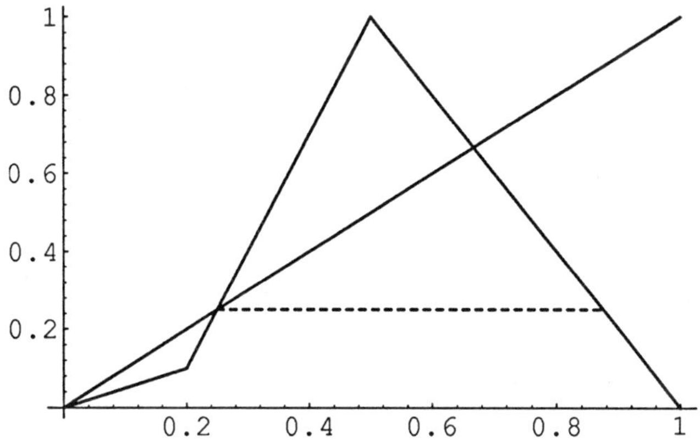

Fig. 6.2.4 The graph of the map of Example 6.2.3 shows that the trajectories of all points x < .25 or x > 7/8 converge to 0.

The origin is an attractive stationary state. The point $x = \frac{4}{11}$ is periodic of period 3. Hence the dynamical system defined by F in [0,1] has infinitely many aperiodic and unstable trajectories, and it is chaotic in the Li-Yorke sense. However, the system is not chaotic in the interval [0, 1] according to DCS.

This remark however, does not exclude the possibility of finding an attractor **A** contained in the interval [1/4, 7/8] such that F is chaotic on **A**.

Other definitions of chaos have been proposed in the literature, besides the one of Li-Yorke and DCS. Some of them involve concepts and facts from measure theory and are therefore beyond the purpose of this book. Other definitions require additional or different properties to be satisfied by F on **A**. As an example we present the one proposed by R.L. Devaney [Devaney, 1987].

Devaney's Definition of Chaotic Systems (DDCS)
Let V be a set. $F : V \rightarrow V$ is said to be **chaotic** on V if

1. F is topologically transitive;
2. periodic points are dense on V;
3. F has sensitive dependence on initial conditions.

Topological transitivity is easily explained when V is a ball. In this case it means that for every pair of balls B_1, B_2, of V there exists an integer n such that

$$F^n(B_1) \cap B_2 \neq \emptyset.$$

By assuming that V is bounded, and by identifying it with an attractor **A** of F, we can compare DCS with DDCS. We did not choose DDCS for the following reasons.

In 1992 Banks J., Brooks J., Cairns G., Davis G., Stacey P. [Banks et al., 1992] proved that property 3 can be removed from Devaney's list, since it is a consequence of the first two.

We already mentioned our reasons for using instability instead of sensitivity and we may add that the definition and use of topological transitivity, beyond those cases where V is a ball, requires Measure Theory. Moreover, the set V should not be finite to avoid degenerate cases.

Therefore, we prefer DCS, although we recognize the merits of the definition proposed by Devaney. In the following example we illustrate how both definitions can be used.

Example 6.2.4 Let B be the map of Example 6.2.2 . We have seen that no orbit of this map is stable. We now prove that there is an orbit $O(x_0)$, which is dense in [0, 1]. Therefore, **C** = **A** = [0, 1].

First we need to illustrate the action of B on the elements of [0, 1] written in base 2.

Let $x < \frac{1}{2}$. Then, in base 2, x = 0.0... and multiplication by 2 will simply move the period by one position to the right.

For example,

$$x = 0.0011 \ (= \tfrac{3}{16}) \text{ implies } 2x = 0.011 \ (= \tfrac{3}{8}).$$

Let $\frac{1}{2} \le x < 1$. In base 2 x is of the form x = 0.1... and B(x) = 2x - 1 will again move the period by one position to the right and replace the first 1 with 0. For example

$$x = \tfrac{5}{8} = 0.101 \text{ becomes } 1.01 - 1 = 0.01 = \tfrac{1}{4}.$$

Since the rationals of the interval [0, 1] are countable we may assume to have listed them in a sequence $\{q_1, q_2, ..., q_n, ...\}$. Form an infinite matrix M according to the following procedure: the entries of row k of M are the beginning blocks of length k of the elements of the sequence $\{q_1, q_2, ..., q_n, ...\}$. For example, if

$$q_1 = .1011..., q_2 = .0111..., q_3 = .1110..., q_4 = .1011..., q_5 = .0100... \text{ etc.},$$

then the first five entries of rows 1, 2, 3, and 4 are

1	0	1	1	0	...
10	01	11	10	01	...
101	011	111	101	010	...
1011	0111	1110	1011	0100	....

Denote the entries of the matrix M by a_{ij} and le $x_0 = .a_{11}a_{21}a_{12}a_{31}a_{22}a_{13}...$.

In the example we obtain

$$x_0 = .11001010111011011111... .$$

All rationals of [0, 1] are limit points of $O(x_0)$, since for every q_i and for every integer k, the number x_0 contains the beginning block of q_i of length k. Therefore, $O(x_0)$ is aperiodic and dense in [0, 1].

The map B is chaotic in [0, 1] according to DCS.

To prove that the system is chaotic in the sense of Devaney we must show that the map is topologically transitive in [0, 1] and the periodic orbits are dense.
The first result is easily established. In fact, let I_1 and I_2 be any two open intervals contained in I. Since the length of I_1 is doubled at each iteration, eventually we must have $B^n(I_1) = I$, and therefore $B^n(I_1) \cap I_2 \neq \emptyset$.

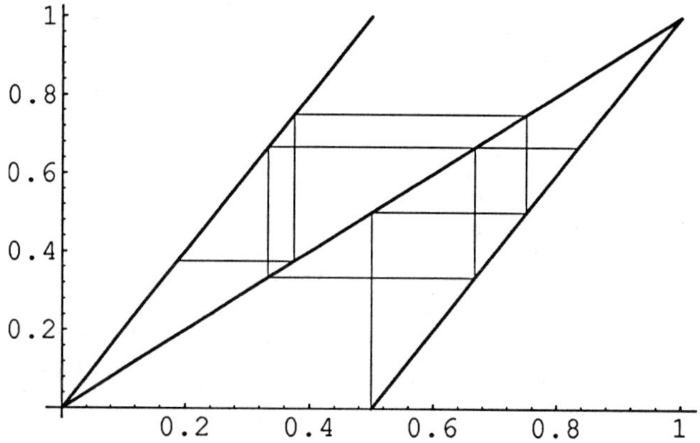

Fig. 6.2.5 The eventually stationary orbit 3/32, 3/16, 3/8, 3/4, 1/2, 1, 0, 0,... and the eventually periodic orbit 5/6, 2/3, 1/3, 2/3, 1/3, ... of B(x) = 2x - [2x] are shown in this graph.

The density of periodic points requires a bit more care. First observe that the trajectory of every rational number of [0, 1], whose decimal expansion in base 2 is finite, is eventually stationary since one of the iterates is 0, while the trajectory of every rational number whose decimal expansion is not finite, is eventually periodic.

For example, the trajectory of $x_0 = \frac{3}{32} = 0.000110000...$ is

$$x_0 = 0.00011 = \frac{3}{32},\ x_1 = 0.0011 = \frac{3}{16},\ x_3 = 0.011 = \frac{3}{8},$$

$$x_4 = 0.11 = \frac{3}{4},\ x_5 = 0.1 = \frac{1}{2},\ x_6 = 0 = x_7 \dots,$$

while the trajectory of $x_0 = 0.1101010...$ (continue in the same manner) $= \frac{5}{6}$ is eventually periodic (of period 2)

$$x_0 = 0.11010... = \frac{5}{6},\ x_1 = 0.1010... = \frac{2}{3},\ x_2 = 0.01010... = \frac{1}{3},$$

$$x_3 = 0.1010.. = \frac{2}{3} = x_1,\ x_4 = x_2,\ \dots .$$

In fact,

$$B(\tfrac{5}{6}) = \tfrac{10}{6} - 1 = \tfrac{2}{3}, \text{ and } B(\tfrac{2}{3}) = \tfrac{4}{3} - 1 = \tfrac{1}{3}\ , B(\tfrac{1}{3}) = \tfrac{2}{3}, \text{ etc.}$$

Hence, we have to show that the rational numbers of $[0, 1]$, whose decimal expansion is infinite and periodic are dense in $[0, 1]$. This part is left as an exercise and it can be accomplished by using the same idea we outlined above to prove that $\mathbf{C} = [0, 1]$.

By the result of Banks et al. the system has sensitive dependence on initial conditions and therefore is chaotic on [0, 1] according to DDCS.

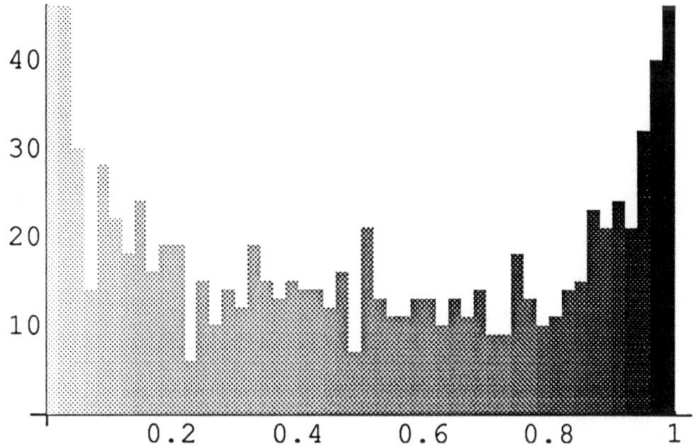

Fig. 6.2.6 The histogram of the logistic map $K(x) = 4x(1 - x)$ provides graphical evidence that the core of the system is the entire interval $[0, 1]$. After dividing $[0, 1]$ into 50 bins and iterating the function K 1500 times starting with $x_0 = .3$, we have placed above every bin a rectangle of height proportional to the number of iterates found inside the bin.

We close this section by proving that the attractor of the system $F(y) = -2y^2 + 1$, with $y \in [-1, 1]$, is the entire interval $[-1, 1]$. To see this we define

$$k : [0, 1] \to [-1, 1] \text{ by } y = k(x) = -\operatorname{Cos} 2\pi x.$$

With

$$B(x) = 2x - [2x] \quad \text{we have} \quad F(k(x)) = k(B(x)).$$

This equality implies

$$F^2(k(x)) = F(F(k(x)) = F(k(B(x)) = k(B(B(x)) = k(B^2(x)),$$

and in general,

$$F^n(y) = k(B^n(x))$$

with $k(x) = y$. Consequently, the limit points of the sequence $\{F^n(y) : n = 0, 1, ...\}$ are images, using k, of the limit points of the sequence $\{B^n(x) : n = 0, 1, ...\}$. Select the initial state x_0 whose orbit $O(x_0)$ is dense in $[0, 1]$. The orbit $O(y_0)$, $y_0 = k(x_0)$, is dense in $[-1, 1]$ and the dynamical system governed by F has the property $\mathbf{C} = \mathbf{A} = [-1, 1]$.

Putting together the above property and the conclusions of Examples 6.2.1 and 6.2.4 we obtain that (we denote the independent variably by x for all maps)

$$F(x) = -2x^2 + 1, \ y \in [-1, 1]\,; \ G(x) = -2|x| + 1, \ y \in [-1, 1];$$

$$K(x) = 4x(1 - x), \ x \in [0, 1]\,; \ B(x) = 2x - 2[x], \ x \in [0, 1];$$

and

$$H(x) = \begin{cases} 2x & 0 \le x < 0.5 \\ 2 - 2x & 0.5 < x \le 1 \end{cases}$$

are chaotic in the stated intervals in the sense of DCS (and of DDCS).

Section 3. Dimension

Efforts to analyze the attractors of dynamical systems, to find similarities among them and differences with other familiar objects of our geometry have been frequently focused on establishing their **dimension.**

This task is easy when the attractor **A** is a finite set of points R or a closed ball $D \subset \mathbf{R}^q$. Using the intuitive idea of dimension we say that $\dim(\mathbf{A}) = 0$ in the first case and $\dim(\mathbf{A}) = q$ in the second case .

However, to capture the peculiarities of some attractors, the standard intuitive definition of dimension is not sufficient and several more elaborate definitions of dimension have been proposed. They can be roughly divided into two groups.

In the **first group** we find those definitions which are simply concerned with the geometry of the set and do not take into account how the set has been generated.

The definitions of the **second group** try to capture the dynamical nature of the attractor looking at how the states of the system are distributed in its various regions, exploring the relations between successive states and measuring the strength of these relations.

We present here two definitions: one from the first and the other from the second group. The first is called *capacity* and the second *correlation dimension.*

1. Capacity and Hausdorff dimension

Technically, the definition of dimension presented here is called **capacity**. However, in all cases examined in this book and almost all cases of interest, the capacity coincides with a more general definition called **Hausdorff dimension.**

Therefore, we shall use also this second terminology. The reader should keep in mind that the method presented here is less general than the original idea of Hausdorff. The fundamental steps are illustrated in the following examples.

Example 6.3.1 Let us consider, in the real line, the segment $S = [0, L]$. The minimum number of segments of length $\frac{1}{2}L$ needed to cover S is obviously 2. The two segments are $[0, \frac{1}{2}L]$, $[\frac{1}{2}L, L]$.

The minimum number of segments of length $\frac{1}{n}L$ needed to cover S is n. In general, given $\varepsilon > 0$, the minimum number of segments of length εL needed to cover S is $1/\varepsilon$. When $1/\varepsilon$ is not an integer we should round it up to the nearest integer.

We see that, denoting by $N(\varepsilon)$ the minimum number of segments of length εL which are needed to cover the interval, we find

$$\lim_{\varepsilon \to 0} \varepsilon N(\varepsilon) = \text{constant} > 0. \tag{6.3.1}$$

From the above formula we derive, when ε is very small,

$$\ln N(\varepsilon) \approx \text{Constant} - \ln \varepsilon .$$

Keeping in mind that $-\ln\varepsilon \to \infty$ as $\varepsilon \to 0$ we obtain

$$\lim_{\varepsilon \to 0} \frac{\ln N(\varepsilon)}{-\ln \varepsilon} = 1. \tag{6.3.2}$$

The limit of (6.3.2) is called the **capacity** (or **Hausdorff dimension**) of the interval $[0, L]$. As expected, this dimension is 1.

Let us now analyze a more complex subset of the real line, called the **Cantor set**.

Example 6.3.2 The Cantor set is generated from the interval $[0, 1]$. The first step of the construction is the removal from $[0, 1]$ of the interval $(1/3, 2/3)$. The second step of the construction is the removal, from what is left, of the intervals $(1/9, 2/9)$ and $(7/9, 8/9)$. The procedure is iterated.

For every integer n we remove the center part from each of the 2^{n-1} intervals left. The length of each removed part is 3^{-n}. What is left at the end of the procedure is the **Cantor set K**. Any point x of $[0, 1]$ is an element of **K** if and only if $x = k3^{-n}$ for some positive integers k and n, $k \le 3^n$.
The total length of all segments we have removed is

$$\frac{1}{3} + \frac{2}{9} + \frac{4}{27} + \ldots + \frac{2^n}{3^{n+1}} + \ldots = \frac{1}{3}\left[1 + \frac{2}{3} + \frac{4}{9} + \ldots + (\tfrac{2}{3})^n + \ldots\right] = \frac{1}{3}\,\frac{1}{1-\frac{2}{3}} = 1.$$

Hence, we can say that **K** has length 0. We could also say that the dimension of **K** is zero. This statement, however, is not our best description of the size of **K**.

Using the idea of Hausdorff, we cover **K** with the minimum number of segments of length 1/3. Two such segments are needed: [0, 1/3] and [2/3, 1]. Then we cover **K** with the minimum number of segments of length 1/9. Four such segments are needed, namely [0, 1/9], [2/9, 1/3], [2/3, 7/9], and [8/9, 1]. In general covering **K** with the minimum number of segments of length 3^{-n} requires 2^n segments. Hence, using the same approach of (6.3.2), we conclude that

$$\lim_{\varepsilon \to 0} \frac{\ln N(\varepsilon)}{-\ln \varepsilon} = \lim_{n \to \infty} \frac{n \ln 2}{n \ln 3} = \frac{\ln 2}{\ln 3} \tag{6.3.3}$$

and we say that **K** has capacity or Hausdorff dimension $\frac{\ln 2}{\ln 3}$.

The idea of capacity, as illustrated by the above examples, is therefore the following. Suppose that A is a subset of $\mathbf{R}^q$. Choose $\varepsilon > 0$ and consider hypercubes (segments in **R**, squares in $\mathbf{R}^2$, cubes in $\mathbf{R}^3$ etc.) of side ε. Denote by $N(\varepsilon)$ the smallest number needed to cover A. Then consider the

$$\lim_{\varepsilon \to 0} \frac{\ln N(\varepsilon)}{-\ln \varepsilon} . \tag{6.3.4}$$

The **capacity** (or **Hausdorff dimension**) of A is the numerical value of this limit.

The Hausdorff dimension of a set A is strongly dependent on the geometry of the set. The possible dynamical properties of A do not play a determinant role in calculating its Hausdorff dimension. Moreover, we need quite a good knowledge of A to estimate $N(\varepsilon)$ accurately. This proves to be non-trivially difficult when dealing with attractors of dynamical systems, since frequently we cannot describe them with equations, inequalities or geometrical procedures. Our next definition of dimension seems to be more suitable for attractors of dynamical systems.

2. **Correlation dimension**

This dimension makes good sense for a set which is generated by a dynamical system $F : D \to D$, where D is some closed ball in $\mathbf{R}^q$. We start with a **time series** of our system, namely a trajectory $\{\mathbf{x}_n : n = 0, 1, ...\}$. For simplicity, we assume $F(\mathbf{x}_0) \neq \mathbf{x}_0$ and $O(\mathbf{x}_0)$ not periodic. What we are computing is the correlation dimension of the set of limit points $L(\mathbf{x}_0)$ of $\{\mathbf{x}_n : n = 0, 1, ...\}$.

Given a positive real number r, we form the so called "correlation integral "

$$(6.3.5) \qquad C(r) = \lim_{n \to \infty} \frac{1}{n^2} \sum_{i \neq j}^{n} H(r - \|\mathbf{x}_i - \mathbf{x}_j\|) ,$$

where

$$(6.3.6) \qquad H(x) = \begin{cases} 0 & x < 0 \\ 1 & x \geq 0 \end{cases}$$

is the heaviside unit-step function. C(r) measures the density of pairs of distinct vectors $\mathbf{x}_i, \mathbf{x}_j$ which are closer than r. In the summation only those indices not larger than n are counted, and the standard Euclidean norm is used. This, however, can be replaced by any other norm.

The **correlation dimension** D_C of $L(\mathbf{x}_0)$ is defined as

$$(6.3.7) \qquad D_C = \lim_{r \to 0} \frac{\ln C(r)}{\ln r}.$$

In practice, when a numerical estimate of D_C is desired, we select a large number N of states of the orbit $O(\mathbf{x}_0) : \mathbf{x}_0, \mathbf{x}_1, ..., \mathbf{x}_N$ and we approximate C(r) for several different values of r.

$$(6.3.8) \qquad C(r) \approx \frac{1}{N^2} \sum_{i \neq j}^{N} H(r - \|\mathbf{x}_i - \mathbf{x}_j\|).$$

Definition (6.3.7) tells us that large values of r are not going to provide any useful information on D_C. At the same time, since we are using only N states, we should avoid values of r which are so small that the estimate of C(r) is meaningless. Thus the values of r should be selected so as to avoid these two extremes.

The ln C(r) is then plotted versus ln r and the slope of the least square line can be accepted as a reasonable estimate of D_C. By using different values of N we can verify how stable the estimate is with respect to increases in the number of states considered.

With suitable changes, the same method can be applied to a dynamical system of which only observable quantities are available, and no governing equations are known with sufficient reliability. In this case one of the observables (usually a scalar) is measured at fixed time intervals and a long time series $\{x_n : n = 0,1,2, \ldots, N\}$ is produced. The measured quantity may represent, among the variables of the system, the one which is easier to estimate or which is more relevant to the dynamical process. It is advisable to discard some terms at the beginning of the sequence to avoid interferences with the transient of the system.

For convenience we rename the elements of the sequence $\{x_n\}$ as follows

(6.3.9) $$x(n) = x_n .$$

We now select three positive integers : m (embedding dimension), L (window gap), and J (shift), and we form the vectors

(6.3.10) $$\begin{cases} v_1 = x(1) \quad x(1+L) \quad \ldots \quad x(1+(m-1)L) \\ v_2 = x(1+J) \quad x(1+J+L) \quad \ldots \quad x(1+J+(m-1)L) \\ \ldots \quad \ldots \quad \ldots \quad \ldots \quad \ldots \\ v_{p+1} = x(1+pJ) \quad x(1+pJ+L) \quad \ldots \quad x(1+pJ+(m-1)L) \\ \ldots \quad \ldots \quad \ldots \quad \ldots \quad \ldots \end{cases}$$

Given a positive real number r, and assuming that the sequence of observables is infinite, the " m-correlation integral " is

(6.3.11) $$C(m, r) = \lim_{N\to\infty} \frac{1}{N^2} \sum_{i\neq j}^{N} H(r - \|v_i - v_j\|) .$$

The **m-correlation dimension** $D_C(m)$ is defined as

(6.3.12) $$D_C(m) = \lim_{r \to 0} \frac{\ln C(m, r)}{\ln r},$$

and the **correlation dimension** D_C is

(6.3.13) $$D_C = \lim_{m\to\infty} D_C(m) .$$

In practice, (6.3.11), (6.3.12) and (6.3.13) must be estimated since we cannot choose N as large as we like, r very small or m very big.
For a given time series of length N and keeping m fixed, we estimate C(m, r) for different values of r like in (6.3.8). The values of $\ln C(m, r)$ are plotted versus $\ln r$ and the slope of the least square line is recorded. This slope should provide a reasonable approximation to $D_C(m)$. As m increases, the slope should converge to an acceptable value for D_C. The parameters L, J and K are used to improve the reliability of the numerical scheme.

Sometimes the constraint $i \neq j$ is replaced by the stronger inequality $|i - j| > K$, where K is an integer larger than 1, to avoid artificial correlates. In fact, dynamical systems arise from modeling real processes and the time step used for the observations may be too small with respect to the time interval the system needs to undergo measurable changes.

The most important merits of the procedure just described, which has been proposed by P. Grassberger and I. Procaccia [Grassberger-Procaccia, 1983] are the following.

1. The computational effort is not prohibitive, although it could be very delicate, and it does not require a full knowledge of the attractor of the system .

2. If no equations of the dynamical process are known and only observables are available, a time series can be constructed using the observable which is easier to measure and it is considered more relevant to the system.

3. In the case when there is uncertainty about the deterministic nature of the dynamical process and there are doubts that random components are predominant in it, the correlation dimension can help in deciding the issue. In fact, in a truly random time series $D_C(m) \to \infty$ as $m \to \infty$. This feature of the real $D_C(m)$ will likely be detected by the numerical scheme.

Example 6.3.3 Consider the dynamical system governed by the function

$$F(x) = \begin{cases} x^2 & \text{if } -3 \leq x \leq 1 \\ 4\sqrt{x} - 3 & \text{if } 1 < x \leq 9 \end{cases}$$

in the interval [-3, 9].

Suppose that we start from the point $x_0 = -2$. We have

$$x_1 = 4,\ x_2 = 5,\ x_3 = 4\sqrt{5} - 3,\ \ldots,\ \ldots$$

and the sequence $\{x_n\}$ is increasing and converges to 9. For every $r > 0$ there is an index $n(r)$ such that all elements of the sequence with index larger than $n(r)$ are closer than r. This implies $C(r) = 1$ and $D_C = 0$, a result we expected since the set of limit points of $O(-2)$ is $L(-2) = \{9\}$.

Example 6.3.4 Consider the dynamical system in $\mathbf{R}^2$ governed by the function

$$F(\mathbf{x}) = F(x, y) = \begin{cases} \dfrac{h(\|\mathbf{x}\|)}{\|\mathbf{x}\|}\,(x\operatorname{Cos}\theta + y\operatorname{Sin}\theta,\ -x\operatorname{Sin}\theta + y\operatorname{Cos}\theta) & \text{if } \|\mathbf{x}\| \neq 0 \\ (0, 0) & \text{if } \mathbf{x} = \mathbf{0} \end{cases}$$

where $h(r) = \sqrt{r}$. Using the standard Euclidean norm, we obtain $\|F(\mathbf{x})\| = \sqrt{\|\mathbf{x}\|}$. Therefore, for every initial condition $\mathbf{x}_0 \neq \mathbf{0}$ we see that $\|\mathbf{x}_n\| \to 1$ as $n \to \infty$. Assume that θ/π is irrational. The same reasoning used at the beginning of this chapter can be applied here to conclude that the dynamical system has only one attractor: the unit circle **A**, which has dimension 1.

We want to verify that the correlation dimension of the attractor is also one. The computation of $C(r)$ can be simplified if we observe that given the form of an orbit $O(\mathbf{x}_0)$ we obtain that $C(r)$ equals the probability that two points on the unit circle are closer than r (in the Euclidean distance). Therefore

$$C(r) = \frac{2}{\pi}\operatorname{Sin}^{-1}\frac{r}{2}$$

and

$$\lim_{r \to 0} \frac{\ln C(r)}{\ln r} = \lim_{r \to 0} \frac{\ln\frac{2}{\pi} + \ln(\operatorname{Sin}^{-1}\frac{r}{2})}{\ln r} = \lim_{r \to 0} \frac{\ln(\operatorname{Sin}^{-1}\frac{r}{2})}{\ln r}.$$

Using l'Hopital's rule, we find

$$\lim_{r \to 0} \frac{\ln(\operatorname{Sin}^{-1}\frac{r}{2})}{\ln r} = 1.$$

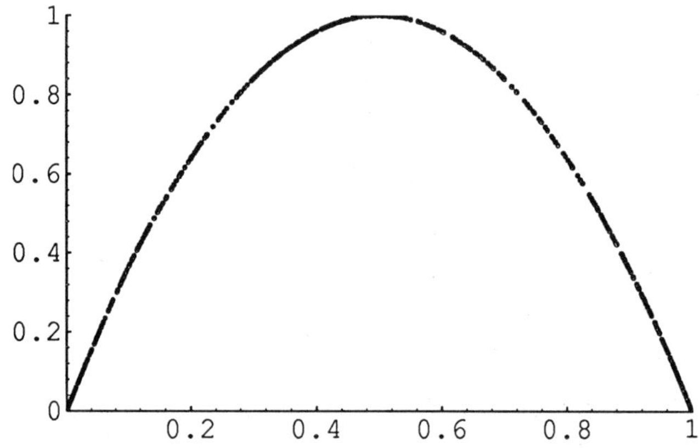

Fig. 6.3.1 The graph has been obtained by plotting the trajectory of $F(x) = 4x(1 - x)$ starting from $x_0 = .3$ versus the same trajectory shifted by one, namely the one starting at .84. The plotted pairs are of the form $(x, F(x))$. The first 200 iterates where discarded and the next 1000 where used for the plot. It is already evident that the trajectory is able to reconstruct the entire graph of F in [0,1]. The trajectory is dense in [0, 1].

In these two examples we have been able to use the first approach since the two dynamical systems are governed by perfectly known deterministic laws. It should be stressed once more that the method proposed by Grassberger-Procaccia is sound also when the system is known to us only through observable quantities.

A final remark is necessary. The definition of correlation dimension and its numerical implementation deals only with the limit set of one trajectory. General conclusions about the overall behavior of the system cannot be derived from the analysis unless we have reason to believe that the orbit is dense in the attractor of the system. This may be the case when the system is chaotic according to DCS, since there is an orbit $O(\mathbf{x}_0)$ which is dense in the attractor **A** (see, for example, Fig. 6.3.1). However, our goal is to use the correlation dimension to detect if a dynamical system is chaotic. Therefore, the numerical investigation should provide some clue either on the instability of all orbits, or on the presence of a dense and aperiodic orbit.

The correlation dimension can be successfully used for both tasks, although it is more suitable for the second. We apply the procedure described in this section to a large number of different time series. Agreement among the numerical estimates we obtain, especially when combined with their non-integer values, provides strong indication that the system is chaotic, or at least has a dense and aperiodic orbit.

The issue of non-integer correlation dimension, however, needs precaution. Since the quantities defined by (6.3.11), (6.3.12) and (6.3.13) are estimated, only numerical values bounded away from any integer can be accepted as evidence of non-integer dimension.

Moreover, we should not forget that there are many chaotic systems for which the correlation dimension of the attractor is an integer, and there are non-chaotic systems with dense and aperiodic orbits (see, for example, the rotation $P(\theta)$). Thus, the numerical analysis should be carried one step further to estimate the Lyapunov exponents, as described in the next section. These numbers provide information on the instability of all orbits and complete the task of checking that both conditions included in DCS are met by the dynamical system.

Another supporting evidence of chaotic behavior is the presence of saddle equilibria, particularly when combined with heteroclinic or homoclinic orbits.

When all information derived from the various sources point in the same direction we can safely assume that the system has indeed the corresponding properties, even though we are unable to establish their presence theoretically.

Section 4. Lyapunov Exponents

As mentioned previously, the attractor **A** of a dynamical system may have integer dimension, even though the system is chaotic on **A**. The following example illustrates such a situation.

Example 6.4.1 Let $K(x) = 4x(1 - x)$, $x \in [0, 1]$. We have seen that the dynamical system governed by K is chaotic in the sense of DCS and the attractor of the system is the entire interval $[0, 1]$. The correlation dimension of the attractor is 1. We found the same result in the non-chaotic system of Example 6.3.4 .

Thus, chaotic dynamical systems cannot be characterized by the fractional dimension of their attractors.

Lyapunov exponents, presented in this section, can be used to test the instability of all orbits, which is one of the properties required by DCS. Again, also in this case, we should keep in mind that there are bounded and non-chaotic systems with no stable orbits.

Therefore, we must use all available strategies before reaching conclusions on the long range behavior of a system.

Like the correlation dimension presented earlier, the Lyapunov exponents are based on the full length of trajectories. They differ from the correlation dimension in one important aspect: instead of simply considering at each step the state of the system in that moment, they take into account its relations with nearby states.

More precisely, given the orbit

$$\mathbf{x}_0,\ \mathbf{x}_1 = F(\mathbf{x}_0),\ \mathbf{x}_2 = F(\mathbf{x}_1),\ \ldots. \tag{6.4.1}$$

where $\mathbf{x}_0$ is a q-dimensional vector, and given a positive integer n, we consider the eigenvalues

$$\lambda_1(\mathbf{x}_0, n), \ldots, \lambda_m(\mathbf{x}_0, n), \qquad m \le q, \tag{6.4.2}$$

of the matrix

(6.4.3) $$F_x(\mathbf{x}_{n-1})\ldots F_x(\mathbf{x}_1)F_x(\mathbf{x}_0) = \frac{d}{dx}F^n(\mathbf{x}_0)\,.$$

Denote by $d_1(\mathbf{x}_0, n), \ldots, d_m(\mathbf{x}_0, n)$ the magnitude of the eigenvalues. Without loss of generality we may assume that

(6.4.4) $$d_1(\mathbf{x}_0, n) \geq d_2(\mathbf{x}_0, n) \geq \ldots \geq d_m(\mathbf{x}_0, n)\,.$$

Define the dominant Lyapunov number $N(\mathbf{x}_0)$ of the trajectory $O(\mathbf{x}_0)$ by

(6.4.5) $$N(\mathbf{x}_0) = \lim_{n\to\infty} \sqrt[n]{d_1(\mathbf{x}_0, n)}$$

where, for each n, the positive real n-th root is taken. The dominant Lyapunov exponent is

(6.4.6) $$\Lambda(\mathbf{x}_0) = \ln N(\mathbf{x}_0)\,,$$

i.e. the natural logarithm of the dominant Lyapunov number $N(\mathbf{x}_0)$.

Example 6.4.2 Let $F(x) = \frac{5}{2}x(1 - x)$ and let $x_0 = 0$. Since $F_x(0) = \frac{5}{2}$ and $F^n(0) = 0$ for all n we have $d(n) = (5/2)^n$. Therefore, the Lyapunov number at 0 (we do not need to say dominant because there is only one) is

$$N(0) = \lim_{n\to\infty} \sqrt[n]{(\tfrac{5}{2})^n} = \frac{5}{2}\,.$$

Thus $\Lambda(0) = \ln 5 - \ln 2$.

Now let $x_0 = \frac{3}{5}$. Notice that $F^n(\frac{3}{5}) = \frac{3}{5}$ for all n and $F_x(\frac{3}{5}) = -\frac{1}{2}$. Therefore $d(n) = (\frac{1}{2})^n$ and

$$N(\tfrac{3}{5}) = \lim_{n\to\infty} \sqrt[n]{(\tfrac{1}{2})^n} = \frac{1}{2}\,.$$

Thus $\Lambda(\frac{3}{5}) = -\ln 2$.

Example 6.4.3 Let $F(x) = 3.2\,x(1 - x)$. The point $x_s = \frac{11}{16}$ is an equilibrium point.

From $F_x(\frac{11}{16}) = -1.2$ we derive $\Lambda(\frac{11}{16}) = \ln 1.2 > 0$. For the periodic orbit of period 2

$$x_0 = \frac{21 + \sqrt{21}}{32} \qquad x_1 = \frac{21 - \sqrt{21}}{32}$$

we find

$$F_x(x_0) = -1 - .2\sqrt{21} \qquad F_x(x_1) = -1 + .2\sqrt{21}\,,$$

and we obtain

$$\frac{d}{dx}F^n(x_0) = \begin{cases} (.16)^m & n = 2m \\ (.16)^m(-1 - .2\sqrt{21}) & n = 2m + 1\,. \end{cases}$$

Consequently,

$$d(x_0, n) = \begin{cases} (.16)^m & n = 2m \\ (.16)^m(1 + .2\sqrt{21}) & n = 2m + 1 \end{cases}$$

and

$$N(x_0) = \lim_{n\to\infty} \sqrt[n]{d(x_0,n)} = .4\,, \quad \Lambda(x_0) = \ln 2 - \ln 5.$$

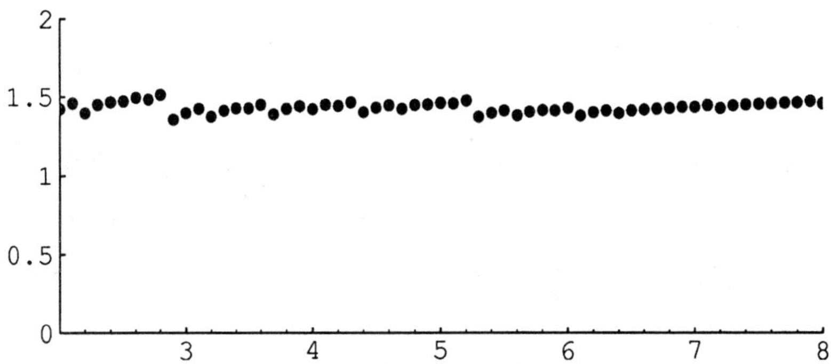

Fig. 6.4.1 The Lyapunov number of the trajectory O(.3) of the map $F(x) = 3.8x(1 - x)$ is numerically estimated with 80 iterations. The last sixty are plotted in this graph. It seems that the sequence converges to 1.5. The orbit appears to be unstable.

As suggested by the previous example, the Lyapunov number is less than 1 when the orbit is stable and larger than 1 when it is unstable. To make this statement more accurate we need however a small adjustment to (6.4.5). To see why this adjustment is needed consider the following example.

Example 6.4.4 Let $K(x) = 4x(1 - x)$ and select the initial condition $x_0 = .25\,(2 - \sqrt{2}\,)$. It can be easily verified that

$$x_1 = 0.5,\ \ x_2 = 0,\ \ x_3 = 0, \dots .$$

Since $K_x(0.5) = 0$, the Lyapunov number of the trajectory $O(x_0)$ is 0. The result suggests that the trajectory is stable, while we have established previously that the dynamical system governed by K has no stable orbits. Moreover, the stationary state $x_s = 0$ is a source since $K_x(0) = 4$.

The presence of the state $x_1 = 0.5$ is responsible for the wrong information provided by the Lyapunov number on the orbit $O(x_0)$. To avoid this inconvenience we propose a modification to (6.4.5).

Given a positive integer k, and $n > k$, consider the eigenvalues

$$(6.4.7)\qquad \lambda_1(\mathbf{x}_0, k, n), \dots, \lambda_m(\mathbf{x}_0, k, n), \qquad m \le q,$$

of the matrix

$$(6.4.8)\qquad F_{\mathbf{x}}(\mathbf{x}_{n-1}) \dots F_{\mathbf{x}}(\mathbf{x}_{k+1}) F_{\mathbf{x}}(\mathbf{x}_k)\,.$$

Let

$$(6.4.9)\qquad d_1(\mathbf{x}_0, k, n) = \max\{\, |\lambda_i(\mathbf{x}_0, k, n)|\,,\ i = 1, 2, \dots, m\}.$$

Define the k-dominant Lyapunov number $N(\mathbf{x}_0, k)$ of the trajectory $O(\mathbf{x}_0)$ by

$$(6.4.10)\qquad N(\mathbf{x}_0, k) = \lim_{n\to\infty} \sqrt[n]{d_1(\mathbf{x}_0, k, n)}\,,$$

where, for each n, the positive real n-th root is taken. The dominant Lyapunov number $N(\mathbf{x}_0)$ and the dominant Lyapunov exponent $\Lambda(\mathbf{x}_0)$ are

$$(6.4.11)\qquad N(\mathbf{x}_0) = \lim_{k\to\infty} N(\mathbf{x}_0, k)\,, \qquad \Lambda(\mathbf{x}_0) = \ln N(\mathbf{x}_0)\,.$$

With this new definition we obtain that both quantities are not affected by any particular state of the trajectory $O(\mathbf{x}_0)$.

For one-dimensional systems the Lyapunov number $N(x_0)$ given by (6.4.11) is the geometric average of the absolute values of the derivative of F along the tails of the orbit $O(x_0)$.
$N(x_0) > 1$ $(\Lambda(x_0) > 0)$ indicates that the orbit is unstable, while $N(x_0) < 1$ $(\Lambda(x_0) < 0)$ indicates that the orbit is stable. The same implications are valid for dynamical systems in $\mathbf{R}^q$, $q > 1$.

Example 6.4.5 With K and x_0 as in the previous example we obtain, for $k \geq 2$,

$$d(x_0, k, n) = [K_x(0)]^{n-k} = 4^{n-k}.$$

Thus

$$N(x_0, k) = \lim_{n \to \infty} \sqrt[n]{4^{n-k}} = 4, \qquad N(x_0) = 4.$$

The effect of $x_1 = \frac{1}{2}$ in the computation of $N(x_0)$ has been eliminated.

Example 6.4.6 Let

$$H(x) = \begin{cases} 2x & 0 \leq x < \frac{1}{2} \\ 2 - 2x & \frac{1}{2} \leq x \leq 1 . \end{cases}$$

Since $|H_x(x)| = 2$ for every $x \neq \frac{1}{2}$ we see that the Lyapunov number of every orbit which does not contain the state $\frac{1}{2}$ is 2. For an orbit $O(x_0)$ which contains $\frac{1}{2}$ we can choose k so large that the state $\frac{1}{2}$ is eliminated from the computation of $N(x_0, k)$, since $H(\frac{1}{2}) = 1$, $H(1) = 0$ and $H(0) = 0$. Hence $N(x_0) = 2$ also in this case.

When the initial state $\mathbf{x}_0$ is a periodic point of prime period $p \geq 1$, the k-dominant Lyapunov number is independent of the integer k, and one can use the simpler method provided by (6.4.5).

Example 6.4.7 Let $F : [0,1] \to [0,1]$ be defined by

$$F(x) = \begin{cases} 4x & 0 \le x \le \frac{1}{4} \\ -3x + \frac{7}{4} & \frac{1}{4} < x \le \frac{1}{2} \\ -\frac{1}{2}x + \frac{1}{2} & \frac{1}{2} < x \le 1 . \end{cases}$$

Then $\frac{1}{6}, \frac{2}{3}$ is a periodic orbit (of prime period 2) and

$$F_x(\tfrac{1}{6}) = 4, \text{ while } |F_x(\tfrac{2}{3})| = \tfrac{1}{2}.$$

We can use (6.4.5) to obtain $d(\frac{1}{6}, n) = 4^q (\frac{1}{2})^p = 2^{2q-p}$ where

$$\begin{cases} q = p = \frac{n}{2} & n \text{ even;} \\ q = p+1 = \frac{n+1}{2} & n \text{ odd .} \end{cases}$$

Since for every number $a > 0$, $\lim_{n\to\infty} \sqrt[n]{a} = 1$, we obtain

$$N(\tfrac{1}{6}) = N(\tfrac{2}{3}) = \sqrt{2}.$$

Thus

$$\Lambda(\tfrac{1}{6}) = \Lambda(\tfrac{2}{3}) = \tfrac{1}{2}\ln 2 > 0.$$

In Fig. 6.4.2 we have plotted the Lyapunov exponents of the one-parameter family of maps $ax(1- x)$ (logistic) when the parameter a ranges in the interval $[3, 4]$. Remarkable is the window of stability around the value $a = 3.83$. In this window the system has a stable periodic orbit of period three.

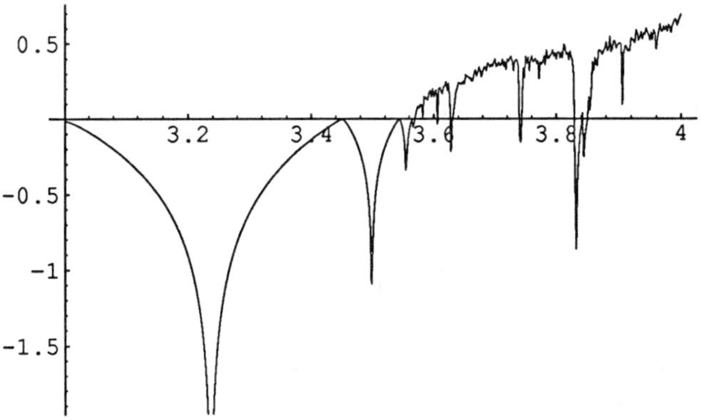

Fig. 6.4.2 The Lyapunov exponent of the logistic map $F(x, a) = ax(1 - x)$ for $a \in [3,4]$.

The map

$$h : [0, 1] \to [0, 1]\ ,\ h(x) = \frac{2}{\pi}\ \mathrm{ArcSin}\sqrt{x}$$

is a conjugacy between the functions K, $H : [0, 1] \to [0, 1]$, $K(x) = 4x(1 - x)$ and

$$H(x) = \begin{cases} 2x & 0 \le x < \frac{1}{2} \\ 2 - 2x & \frac{1}{2} \le x \le 1\ . \end{cases}$$

It is natural to ask if the conjugacy h can be used to evaluate the Lyapunov number of an orbit $O(x_0)$ of K using the equality

$$(6.4.12) \qquad K^n(x_0) = h^{-1}H^n h(x_0)$$

which is valid for every n, and the result $N(h(x_0)) = 2$ established in Example 6.4.6 for H. The answer to this question is affirmative and we can say that for the dynamical system governed by K we have $N(x_0) = 2$ as well.

There are exceptions. For example, the Lyapunov number of the stationary orbit $O(0)$ is 4 when the dynamical system is governed by K, and it is 2 when it is governed by H. Notice that h is not differentiable at 0.

There is another question connected to the one we just asked. How does $N(x)$ behave as x ranges over the interval I where the action of a function $T : I \to I$ is taking place ?

The study of both questions requires more background than this book assumes. Using ideas from measure theory one can establish that under certain conditions, which are met by many dynamical systems, the following two statements hold.

1. *Given* $F : I \to I$ *the Lyapunov number* $N(x)$ *is independent of* x *for almost all* $x \in I$.

2. *Given* $F : I \to I, G : J \to J$ *and a conjugacy* $h : I \to J$, *we have* $N(x) = N(h(x))$ *for almost all* $x \in I$.

The "almost all" restriction tells us that exceptions should be expected. The size of the subset $R \subset I$ of almost all x for which the statements hold can be quantified by saying that with respect to some probability measure P we have $P(R) = P(I) = 1$. In other words, the probability of finding an initial state $x_0 \in I$, such that $x_0 \notin R$ is 0.

Statements 1 and 2 can be extended to dynamical systems in $\mathbf{R}^q$, $q > 1$. Moreover the process which leads to the dominant Lyapunov number using $d_1(\mathbf{x}_0, k, n)$ can be repeated for every $d_i(\mathbf{x}_0, k, n)$, although d_1 is the most important from a dynamical point of view. In the one-dimensional case we have omitted the qualification "dominant" since the derivative has only one "eigenvalue".

When the dimension q of the system is large the search for $d_1(\mathbf{x}_0, k, n)$ is difficult. Since d_1 is the spectral radius of the matrix $A(\mathbf{x}_0, k, n) = F_\mathbf{x}(\mathbf{x}_{n-1})...F_\mathbf{x}(\mathbf{x}_k)$, the so-called **power method** can be used to estimate it. Assume that

1. all eigenvalues of $M = A(\mathbf{x}_0, k, n)$ are semisimple, and

2. $|\lambda_1(\mathbf{x}_0, k, n)| > |\lambda_2(\mathbf{x}_0, k, n)| \geq ... \geq |\lambda_m(\mathbf{x}_0, k, n)|$.

Then, for almost all vectors $\mathbf{v} \in \mathbf{R}^q$, the sequence

$$(6.4.13) \qquad \frac{M\mathbf{u}_p \bullet \mathbf{u}_p}{\|\mathbf{u}_p\|^2}$$

converges to $\rho(M)$, where the elements of the sequence are defined as follows:

$$(6.4.14) \qquad \mathbf{u}_0 = \mathbf{v},\ \mathbf{u}_1 = \frac{1}{r_0} M\mathbf{u}_0,\ \mathbf{u}_2 = \frac{1}{r_1} M\mathbf{u}_1 \ldots$$

with $r_p = \max\{\ |M\mathbf{u}_p \bullet \mathbf{e}_i| : i = 1, 2, ..., q\}$.

Example 6.4.8 The eigenvalues of the matrix

$$M = \begin{pmatrix} 2 & 2 \\ 1 & 3 \end{pmatrix}$$

are $\lambda_1 = 1$ and $\lambda_2 = 4$. Hence $\rho(M) = 4$.
We select $\mathbf{v} = (1, 0)$ and apply the power method to obtain

$$\mathbf{u}_1 = (1, 0.5),\ \mathbf{u}_2 = (1, \frac{5}{6}),\ \mathbf{u}_3 = (1, \frac{21}{22}),\ M\mathbf{u}_3 = (\frac{43}{11}, \frac{85}{22}).$$

The approximate value of $\rho(M)$ after these four iterations is 3.98.

Computing $N(k, \mathbf{x}_0)$ and $N(\mathbf{x}_0)$ is time consuming and margins of uncertainty cannot be eliminated due to the presence of the limits (6.4.5), (6.4.10) and (6.4.11), which, for obvious reasons, can only be estimated in any numerical scheme. Moreover, when q is large, these problems are regularly compounded by the task of finding the spectral radius of the relevant matrices. Therefore, only after extensive and careful investigation should one feel comfortable with deriving conclusions on the instability of the dynamical process.

We close this chapter by mentioning once more that the final goal of the numerical analysis performed to determine the correlation dimension and the Lyapunov exponents is to detect the chaotic behavior of the system. When the results from several different orbits (time series) are in agreement with each other and suggest indecomposability and instability, particularly in the presence of saddle equilibria and homoclinic or heteroclinic orbits, we can label the system as chaotic, at least pending a theoretical proof of the contrary.

Problems

Section 1.

6.1.1 Find a periodic orbit of prime period 3 of the map

$$F(x) = \begin{cases} \frac{x}{2} & \text{if} \quad 0 \leq x \leq 0.2 \\ 3x - \frac{1}{2} & \text{if} \quad 0.2 < x \leq \frac{1}{2} \\ 2 - 2x & \text{if} \quad \frac{1}{2} < x \leq 1 . \end{cases}$$

6.1.2 Investigate the domain of attraction of the origin for the map of Problem 6.1.1 .

6.1.3 Let $F(x) = \frac{1}{4}x^2 - 1$. Find an interval with non empty interior which is mapped into itself by F and find the core and the attractor of F in the interval .

6.1.4 Prove that the core of a dynamical system $F : [0, 1] \to [0, 1]$, where F is continuous cannot be finite if the system has a periodic orbit of period 6.

6.1.5 Is it true that the limit set of an asymptotically periodic orbit is always a building block of the core ?

6.1.6 a) Let $F(x) = 2x(1 - x)$. Verify that $x_s = \frac{1}{2}$ is a stationary state of F.

b) Choose $x_0 = \frac{1}{2} + h$ with $|h| < \frac{1}{2}$. Verify that $x_n \to \frac{1}{2}$ as $n \to \infty$.

Find the core of F.

Section 2.

6.2.1 Let $F(\mathbf{x}) = F(x, y) = (x\text{Cos}\,\theta - y\text{Sin}\,\theta,\ x\text{Sin}\,\theta + y\text{Cos}\,\theta)$. Using the Euclidean norm in $\mathbf{R}^2$ prove that $\|F(x, y)\| = \|(x, y)\|$.

6.2.2 Let F be the baker transformation of the unit interval [0, 1]. Given $y_0 \in [0, 1]$, $y_0 = .a_1a_2a_3a_4...$ prove that the subsequence $\{x_0, x_1, x_3, x_6, x_{10}, x_{15}, x_{21}, ...\}$ of the orbit starting from $x_0 = .a_1a_1a_2a_1a_2a_3a_1..a_4a_1..a_5a_1..a_6..$ converges to y_0.

6.2.3 Show that the rational numbers of the interval [0, 1] whose decimal expansion is periodic and it contains infinitely many digits equal to 1 are dense in [0, 1].

6.2.4 With
$B : [0, 1] \to [0, 1]$, $F : [-1, 1] \to [-1, 1]$ and $k : [0, 1] \to [-1, 1]$ defined by

$$B(x) = 2x - [2x] \quad , F(y) = -2y^2 + 1 \quad \text{and } k(x) = -\text{Cos}2\pi x$$

prove that $F(k(x)) = k(B(x))$.
Verify that the above equality implies $F^n(y_0) = k(B^n(x_0))$ with $k(x_0) = y_0$.

6.2.5 From the results of the previous problem and knowing that the core of B is the entire interval [0, 1], derive that the core of the dynamical system defined by the map F in [-1, 1] is the entire interval.

6.2.6 Let $F(x) = xe^{x^3 - x}$. Verify that F maps the interval [-3, 1] into itself.
Verify also that -1 , 0 and 1 are stationary states of F.

6.2.7 Let $F(x) = xe^{x^3 - x}$. Verify that for every x_0,
$0 < x_0 < 1$, we have $x_n \to 0$ as $n \to \infty$.
Verify that 0 is not stable.

6.2.8 Use Theorem 6.2.1 to prove the equality $h(F(x)) = G(h(x))$ where the functions F, $G, h : [-1, 1] \to [-1, 1]$ are defined by

$$F(x) = -2x^2+1, \quad G(x) = -2|x| +1 \quad \text{and} \quad h(x) = \frac{4}{\pi}\text{ArcSin}\left(\frac{1+x}{2}\right)^{0.5} - 1 .$$

6.2.9 Let F and G be as in the previous problem. Let x_s be a stationary state and x_0 be a periodic point of prime period p of F. Using the chain rule show that

$$F_x(x_s) = G_x(h(x_s)) \quad \text{and} \quad (F^p)_x(x_0) = (G^p)_x(h(x_0)),$$

with suitable assumptions on $h_x(x_s)$ and $h_x(x_0)$.
From these results derive that F does not have any stable fixed points or periodic orbits.

Section 3.

6.3.1 Estimate numerically the correlation dimension of the attractor of the dynamical system governed by the map $F(x) = 3.8\,x(1-x)$.

6.3.2 Estimate numerically the correlation dimension of the attractor of the dynamical system governed by the piecewise linear map

$$F(x) = \begin{cases} \dfrac{3x}{2} & \text{if} \quad 0 \le x < 0.5 \\ \dfrac{3}{2} - \dfrac{3x}{2} & \text{if} \quad 0.5 \le x \le 1 . \end{cases}$$

6.3.3 We now construct a set which is called the Sierpinski triangle. The first step of our construction is to take an equilateral triangle T, join the middle points of the three sides with line segments and remove from T the equilateral triangle so obtained.

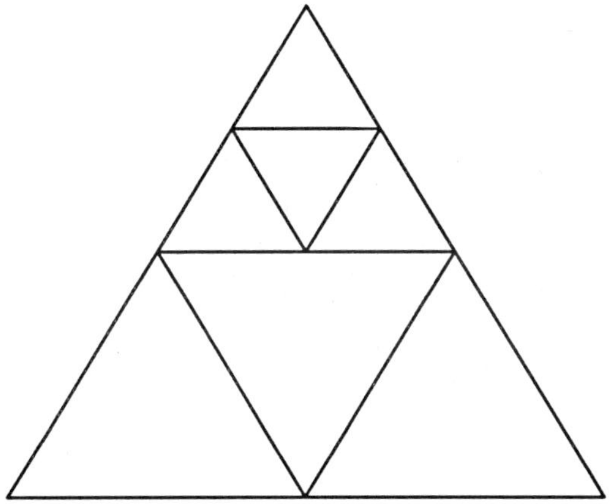

Fig. 6.5.1 The Sierpinski triangle.

The second step is the repetition of the first step, performed on the three equilateral triangles left in T after the removal of the central part. The repetition of this process for every integer n produces, at the end, a geometric figure in the plane which is called the Sierpinski triangle. Find the capacity of the Sierpinski triangle.

6.3.4 Again start with an equilateral triangle T. Remove the center third of each side and substitute it as in the figure below. Keep repeating this procedure. The limiting perimeter is called Koch snowflake. Prove that not only its total length is infinite, but the distance, along the perimeter, between any pair of points on the snowflake is infinite. Prove that the capacity of the snowflake is $\frac{\ln 4}{\ln 3}$.

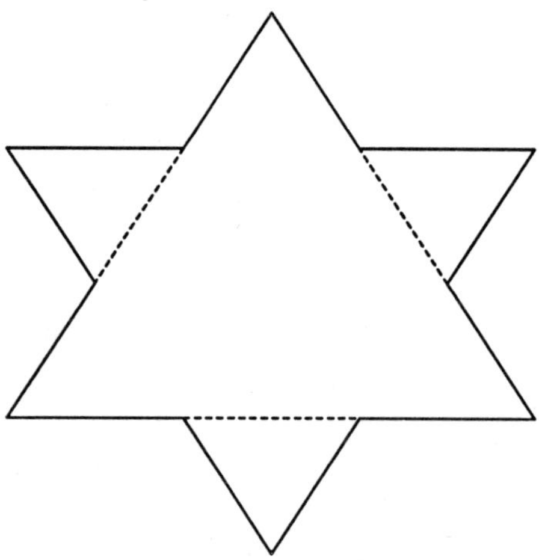

Fig. 6.5.2 The Koch snowflake.

Section 4.

6.4.1 Let F be as in Example 6.4.7. Verify that $\frac{15}{88}, \frac{15}{22}$ is a periodic orbit (of period 2).

Show that for every positive integer n we can write

$$d(n) = 4^q \left(\frac{2}{3}\right)^p$$

where

$$\begin{cases} q = p = \frac{n}{2} & \text{n even} \\ q = p+1 = \frac{n+1}{2} & \text{n odd} . \end{cases}$$

Deduce, from these equalities, the result $N(\frac{15}{88}) = N(\frac{15}{22}) = \frac{2\sqrt{2}}{\sqrt{3}}$.

6.4.2 Let $B(x) = 2x - [2x]$. Verify that $x_0 = \frac{1}{7}$ is a periodic point of prime period 3. Find the Lyapunov number of the orbit.

6.4.3 Let B be as in the previous problem. Consider the orbit with initial state $\frac{1}{28}$. Find the Lyapunov k-number of this orbit. Find the Lyapunov number of the orbit.

6.4.4 Let F be a dynamical system in the real line and assume that $x_0, x_1, ..., x_{p-1}$ is a periodic orbit of period p of F. Prove that $N(k, x_0) = N(x_0)$ for all k.

6.4.5 Let F be a dynamical system in the real line and assume that $x_0, x_1, ..., x_{p-1}$ is a periodic orbit of period p. Prove that $N(x_0) = N(x_1) = ... = N(x_{p-1})$.

6.4.6 Let $F(x) = x^2 - 1$. Find N(0).

6.4.7 Let $F(x) = \frac{1}{4}x^2 - 1$. Find N(0, 0) and N(0).

6.4.8 Let $F(x) = \frac{1}{4}x^2 - 1$. Find $N(1, 2 + 2\sqrt{2})$ and $N(2 + 2\sqrt{2})$.

6.4.9 Let $H(x) = 1 - 2x^2$ and $x_0 = \frac{1}{3}$. Test numerically that the trajectory starting from x_0 is not periodic. Find $N(x_0)$ using the conjugacy between H and the the map $G(x) = 1 - 2|x|$.

6.4.10 Let $M = \begin{pmatrix} -3 & -2 & 1 \\ -2 & -6 & 2 \\ 1 & 2 & -3 \end{pmatrix}$.

Estimate the spectral radius of M using the power method. Then find the spectral radius of M and compare the two results.

CHAPTER 7

ANALYSIS OF FOUR DYNAMICAL SYSTEMS

SUMMARY

In this last chapter we analyze the dynamical systems which were introduced in the second section of chapter 1. We are not presenting a comprehensive and detailed study of the systems, since such a task would require a book on its own. Rather we are focusing on certain features of each system, just to show how a more extensive investigation can be designed and implemented. Sections 1, 2, 3 and 4 are respectively devoted to the one-, two-, three- and multi-dimensional model.

Section 1. A Blood-Cell Population Model

As mentioned in the first chapter, the population of red blood cells at time n+1 can be written as

$$(7.1.1) \qquad x_{n+1} = x_n - d_n + p_n ,$$

where d_n and p_n account respectively for the number of cells destroyed and released into the blood stream during the time interval $[n, n+1]$. The two functions d_n and p_n depend on x_n and possibly on the population at some previous state x_{n-k}. In this model, however, we shall consider the simple case in which d_n and p_n depend only on x_n.

The form of this dependence can obviously be a matter of debate, and there are several forms which are suitable, and none of them can claim to be the best. The form we propose is due to A. Lasota [Lasota, 1977]. He assumed

$$(7.1.2) \qquad d_n = ax_n, \qquad p_n = b(x_n)^r e^{-sx_n}.$$

with $0 < a \leq 1$, and $b, r, s > 0$ to be determined. Our model takes the form

$$(7.1.3) \qquad x_{n+1} = (1 - a)x_n + b(x_n)^r e^{-sx_n} ,$$

and the function governing the dynamical process is

$$(7.1.4) \qquad F(x, \mathbf{a}) = (1 - a)x + bx^r e^{-sx}.$$

Notice that $F(0, \mathbf{a}) = 0$ for every $\mathbf{a}$. Thus $x_s = 0$ is always a stationary point. We expect for this point to be an attractor since, from a clinical point of view, an individual cannot recover when the population of blood cells falls below a certain critical level. Since $F_x(x, \mathbf{a}) = 1 - a + bx^{r-1} e^{-sx}(r - sx)$ we see that the origin is an attractor if the parameter $r > 1$. In this case $F_x(0, \mathbf{a}) = 1 - a < 1$.

Experimental evidence (see [Gearhart-Martelli, 1990]) suggests the presence of a maximum point x_M such that the corresponding maximum value $F(x_M, \mathbf{a})$ is larger than x_M. To account for this feature we first set the derivative of F equal 0. We obtain

(7.1.5) $$bx^r e^{-sx} = x\frac{1 - a}{sx - r}.$$

Hence, at the maximum point x_M, we must have

(7.1.6) $$r < sx_M$$

since $1 - a > 0$. From the required inequality

(7.1.7) $$F(x_M, \mathbf{a}) \geq x_M ,$$

and from (7.1.3), (7.1.5) and (7.1.6) we derive

(7.1.8) $$r < sx_M \leq r + \frac{1 - a}{a}.$$

At this point the best thing to do is to use reliable data and estimates to determine values of the parameters b, r and s. The reader can consult the paper of W. Gearhart and M. Martelli [Gearhart-Martelli, 1990] for this approach. Suggested values for some cases are $r = 8$, $s = 16$ and $b = 1.1\times10^6$.

It can be shown that for every value of $a > 0$ the dynamical system governed by

(7.1.9) $$F(x, a) = (1 - a)x + 1.1\times10^6x^8e^{-16x}$$

has three equilibrium points

$$x_{s1} = 0 < x_{s2} < x_{s3} .$$

We have seen previously that 0 is always a sink. More precisely, every orbit $O(x)$ such that $x < x_{s2}$ converges to 0. The second equilibrium x_{s2} is always a source and there exists $a_0 \in (0,1)$ such that the third equilibrium is a sink for $a < a_0$ and a source for $a > a_0$.

We now determine a_0. In Fig. 7.1.1 we present the graph of the function F for two different values of a, namely $a = 0.2$ and $a = 0.78$ (dashed). The derivative of F at x_{s3} is negative in both cases and it seems that in the case $a = 0.2$ the absolute value of the derivative is smaller than 1, while for $a = 0.78$ is larger than 1. Thus $a_0 \in (.2, .78)$. The stability is lost for those values of a for which the derivative of F at x_{s3} is smaller than -1. Since x_{s3} satisfies the equation

$$(7.1.10) \qquad 1.1\times10^6\,(x_{s3})^7 e^{-16x_{s3}} = a\,,$$

and we must have

$$(7.1.11) \qquad -1 = F_x(x_{s3}, a) = 1 + 7a - 16ax_{s3}\,,$$

we obtain $x_{s3} = \dfrac{2+7a}{16a}$. Substituting into (7.1.10) gives

$$(7.1.12) \qquad 1.1\times10^6\left(\frac{2+7a}{16a}\right)^7 e^{-\frac{2+7a}{a}} = a\,,$$

or

$$(7.1.13) \qquad 1.1\times10^6(2+7a)^7\, e^{-\frac{2+7a}{a}} = 2^{28}a^8\,.$$

Taking the logarithms on both sides we arrive at the equation

$$a\ln 11 + 5a\ln 10 + 7a\ln(2+7a) - 2 - 7a - 28a\ln 2 - 8a\ln a = 0.$$

The solution we are interested in is $a = .262727\ldots$. From (7.1.9), the equilibrium point is $x_s = .913279\ldots$.

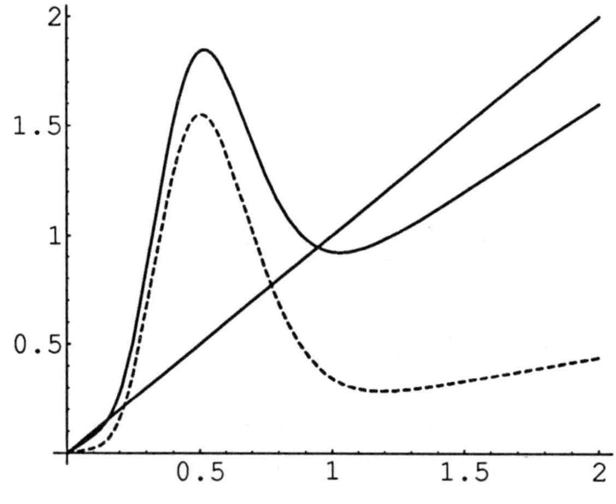

Fig. 7.1.1 The graph of the function $F(x, a) = (1 - a)x + 1.1\times10^6 x^8 e^{-16x}$ for $a = 0.2$ and $a = 0.78$.

In Fig. 7.1.2 we use the value a = .3 . We see that a periodic orbit of period 2 is arising, thus showing that the parameter is crossing a period doubling bifurcation point. As mentioned in the paper of Gearhart-Martelli, the appearance of this periodic orbit of period 2 is in agreement with laboratory findings.

For the value a = .78 we present the graph of F and the graph of its third iterate (see Fig. 7.1.3). We find two periodic orbits of period 3. Hence, for this value of the parameter, the system is chaotic in the sense of Li and Yorke.

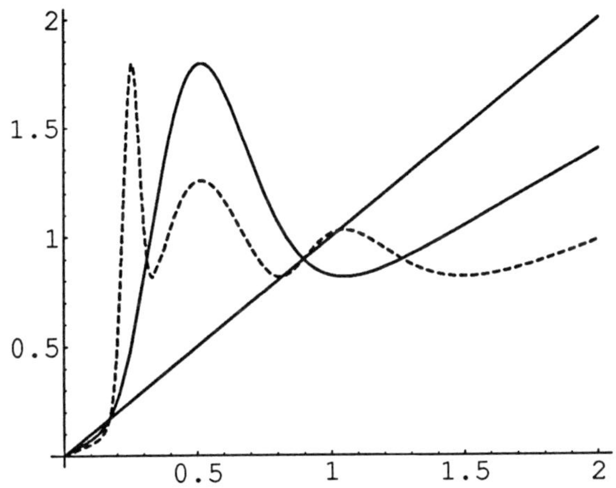

Fig. 7.1.2 Graph of F and F^2 for a = 0.3 .

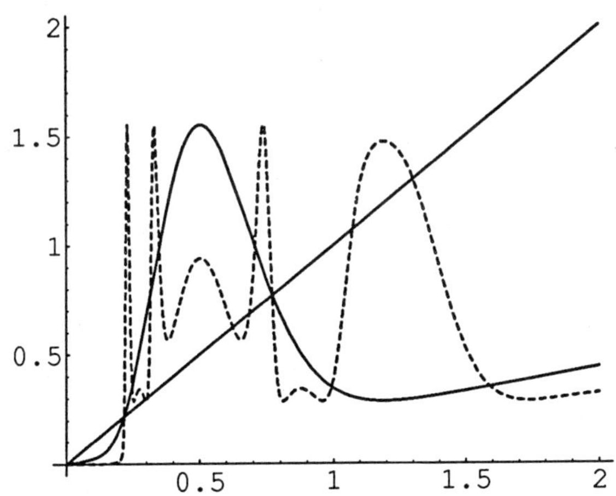

Fig. 7.1.3 Graph of F and F^3 for a = 0.78 .

Also this behavior seems to be supported by laboratory findings. A test on the Lyapunov exponents of the system reveals that the trajectories are unstable for values of a close to .8 .

For example the value $a = .8$ gives the following successive approximations to the Lyapunov exponent along the trajectory starting at $x = .8$.

901	0.43703
902	0.438573
903	0.438806
904	0.436471
905	0.438063

On the left we see the number of iterates, and on the right, the corresponding value of the numerical estimate of the Lyapunov exponent along the trajectory. Starting from 1.1 we get the following table:

901	0.409193
902	0.410843
903	0.411094
904	0.410868
905	0.408586

The agreement is pretty remarkable, since we have only used 900 iterations. It suggests that the system does not have any stable orbits.

For values of a close to 1 we have $F(F(x_M,a), a) < x_{s2}$. This might be expected. When $a \approx 1$ the body tries to compensate for the high destruction rate by increasing the production to its maximum . Consequently, a very large number of cells enters the blood stream, causing an excessive accumulation. This "crowding" is sensed by the automatic controls of the cells level and during the next time interval only a small number of cells is released. Since the destruction rate is unchanged, the population falls below the critical level x_{s2} and the orbit goes to 0.

A more detailed analysis of this model can be found in the article of Gearhart-Martelli.

Section 2. Predator-prey Model of Competition between two Species

The model was proposed originally in the form of a system of differential equations

$$(7.2.1) \qquad \begin{cases} x' = ax - bxy \\ y' = -cy + dxy . \end{cases}$$

where a, b, c, d are positive constants. The variables x and y represent prey and predators respectively. It is assumed, although not very realistically, that in absence of y, the prey population grows exponentially. In fact, from

$$x' = ax \quad \text{we derive} \quad x(t) = x_0 e^{at}.$$

Thus $x(t) \to \infty$ as $t \to \infty$.

In absence of prey, the predators will not survive, since

$$y' = -cy \quad \text{implies} \quad y(t) = y_0 e^{-ct}.$$

Thus $y(t) \to 0$ as $t \to \infty$. This result is in line with expectations.

Prey-predator encounters are obviously favorable to predators (+dxy) and fatal to prey (-bxy).

The presence of outside harvesting activity can be added to the model by writing

$$(7.2.2) \qquad \begin{cases} x' = ax - bxy - \varepsilon x = (a - \varepsilon)x - bxy \\ y' = -cy + dxy - \varepsilon y = -(c + \varepsilon)y + dxy . \end{cases}$$

The equilibrium points of systems (7.2.1) and (7.2.2) are obtained by setting

$$x' = y' = 0.$$

The origin satisfies this condition for both systems. A second equilibrium point is given by

$$(7.2.3) \qquad \left(\frac{c}{d}, \frac{a}{b}\right) \text{ for } (7.2.1) \quad \left(\frac{c+\varepsilon}{d}, \frac{a-\varepsilon}{b}\right) \text{ for } (7.2.2).$$

We see that the presence of a moderate harvesting activity favors the prey population. Their equilibrium value increases, while the predators' equilibrium value decreases. The model was proposed by the Italian mathematician Vito Volterra [Volterra, 1931] to explain to his puzzled friend and son-in-law, the biologist Umberto D'Ancona, the dramatic increase in the population of selachians in the Adriatic Sea during World War I. Volterra argued that the absence of sustained fishing activity during the war changed the normal equilibrium situation. The prey's average value decreased and the predators' increased.

System (7.2.1) cannot be solved by explicitly finding x(t) and y(t). One can solve it numerically. However, the scheme adopted is extremely important. In Chapter 1 we used the one-step Euler method. This approach is acceptable if a small time-step is used and we are not interested in the long range evolution of the system. We arrive at

$$(7.2.4) \qquad \begin{cases} x_{n+1} = (1 + ah)x_n - bhx_n y_n \\ y_{n+1} = (1 - ch)y_n + dhx_n y_n . \end{cases}$$

The stationary states of our system are again (0, 0) and $(\frac{c}{d}, \frac{a}{b})$. We want to analyze their stability. Since $F((x, y), \mathbf{a}) = ((1 + ah)x - bhxy, (1 - ch)y + dhxy)$ we obtain

$$F_x((x, y), \mathbf{a}) = \begin{pmatrix} 1+ah-bhy & -bhx \\ dhy & 1-ch+dhx \end{pmatrix}.$$

Hence,

$$F_x((0, 0), \mathbf{a}) = \begin{pmatrix} 1+ah & 0 \\ 0 & 1-ch \end{pmatrix}$$

whose eigenvalues are $1 + ah$, $1 - ch$. The first is larger than 1 and the second is smaller than 1. Therefore the origin is a saddle.

Similarly, we obtain

$$F_x((\frac{c}{d}, \frac{a}{b}), \mathbf{a}) = \begin{pmatrix} 1 & \frac{-bch}{d} \\ \frac{adh}{b} & 1 \end{pmatrix}$$

whose eigenvalues are $1 + ih\sqrt{ac}$ and $1 - ih\sqrt{ac}$. This second equilibrium is a repeller. To visualize the behavior of the systems let us choose the numerical values $1 + ah = 1.2$, $bh = .2$, $1 - ch = .8$, $dh = .2$. The equilibrium point is (1, 1).

Starting at (1.3, 1) the two populations (see Fig. 7.2.1) display an out-of-phase oscillatory behavior and the amplitudes of the oscillations are increasing. This behavior is expected, since the eigenvalues of $F_x(x_s, \mathbf{a})$ are complex and their modulus is larger than 1.

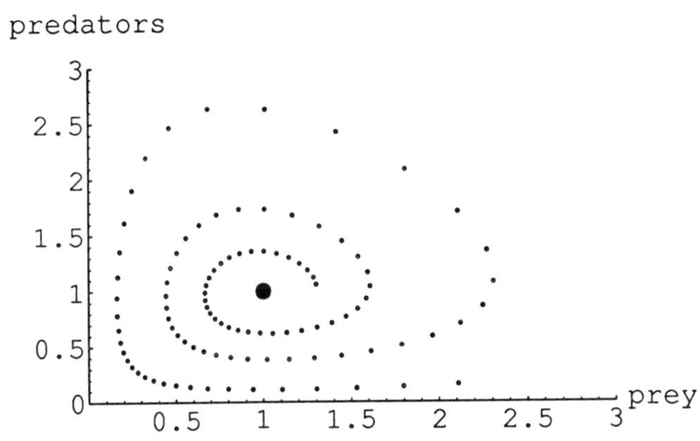

Fig. 7.2.1 The equilibrium point and the trajectory starting at (1.3, 1).

In the continuous system it can be shown that the orbit starting from (1.3, 1) is periodic. This fact underlines the importance of a careful qualitative analysis of a system before using numerical methods for investigating its evolution patterns.

The model of Volterra is based on the unrealistic assumption of exponential growth of the prey in absence of predators. Improvements have been proposed by requiring, for example, that the prey's growth be governed by a logistic map of the form $ax(1 - x)$, and the predators' by $-cy(1 + y)$. Consequently, our discrete one-step Euler version of the new model is

$$(7.2.5) \qquad \begin{cases} x_{n+1} = (1 + ah)x_n - ah(x_n)^2 - bhx_ny_n \\ y_{n+1} = (1 - ch)y_n - ch(y_n)^2 + dhx_ny_n \end{cases}$$

where the time step h should be considered a very small parameter.

The origin is again a stationary state and with

$$T((x, y), \mathbf{a}) = ((1 + ah)x - ahx^2 - bhxy, (1 - ch)y - chy^2 + dhxy),$$

we find

$$T_{\mathbf{x}}((0, 0), \mathbf{a}) = \begin{pmatrix} 1+ha & 0 \\ 0 & 1-hc \end{pmatrix},$$

whose eigenvalues are $1 + ha$ and $1 - hc$. Therefore, the origin is again a saddle.

There are three more equilibria. One has coordinates $(0, -1)$ and it is therefore not suitable for our purposes. Another has coordinates $(1, 0)$. We shall denote it by $\mathbf{x}_2$. Assuming that $c < d$ we have a third equilibrium point

$$\mathbf{x}_3 = \left(\frac{ac + bc}{ac + bd}, \frac{ad - ac}{ac + bd}\right).$$

The stationary state $(1, 0)$ is a saddle since

$$T_{\mathbf{x}}((1,0), \mathbf{a}) = \begin{pmatrix} 1-ha & -bh \\ 0 & 1-hc+dh \end{pmatrix},$$

whose eigenvalues are $1 - ha$ and $1 + h(d - c)$. The last eigenvalue is larger than 1 (since $d > c$).

The analysis of the stability of the third equilibrium point is a bit more complicated. Notice that $T = I + hG$, where

$$G(\mathbf{x}, \mathbf{a}) = (ax(1 - x) - bxy, -cy(1 + y) + dxy).$$

Therefore the eigenvalues of the derivative of T are found by adding 1 to the eigenvalues of G multiplied by h. We have

$$G_{\mathbf{x}}(\mathbf{x}_3, \mathbf{a}) = \frac{1}{ac+bd}\begin{pmatrix} -ac(a+b) & -bc(a+b) \\ ad(d-c) & -ac(d-c) \end{pmatrix}.$$

Since the trace of this matrix is negative and the determinant is positive, its eigenvalues have negative real part. Therefore, for h sufficiently small, the modulus of the eigenvalues of $T_x(\mathbf{x}_2, \mathbf{a})$ is smaller than 1 and the equilibrium is an attractor.

One may find reasons for criticism also in the logistic model of the predator-prey competition, arguing that the origin should be an attractor, since both populations should become extinct if they reach a certain minimum level. The following predator-prey model presents this feature.

$$\text{(7.2.6)} \qquad \begin{cases} x_{n+1} = rx_n(1 - x_n - y_n) \\ y_{n+1} = rx_ny_n \end{cases} \qquad \text{or} \quad \mathbf{x}_{n+1} = F(\mathbf{x}_n, r),$$

with $F(\mathbf{x}, r) = F((x, y), r) = (rx(1 - x - y), rxy)$ and $r \in (0, 4]$. The stationary states of this system are

$$\mathbf{x}_1(r) = (0, 0),\ \mathbf{x}_2(r) = (1 - r^{-1}, 0), \mathbf{x}_3(r) = (r^{-1}, 1 - 2r^{-1}).$$

The first one is an attractor for $r < 1$. At this value of the parameter we have an exchange of stability between the branches $\mathbf{x}_1(r)$ and $\mathbf{x}_2(r)$. The equilibria of the first branch become saddles and the equilibria of the second branch become sinks. Notice that the stationary states of the second branch make sense only for $r > 1$ and at $r = 1$ we have

$$\mathbf{x}_2(1) = (0, 0) = \mathbf{x}_1(1).$$

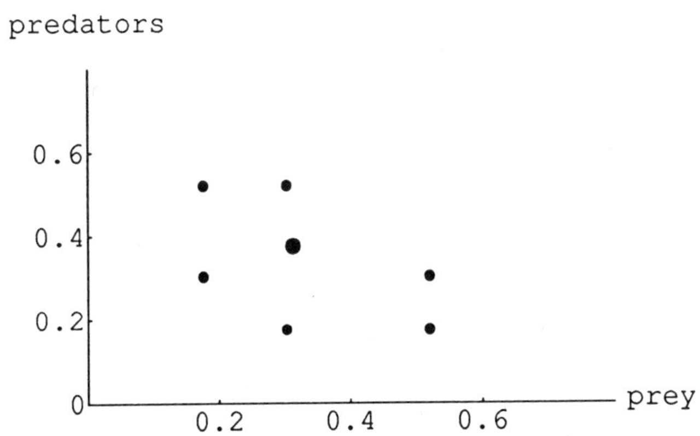

Fig. 7.2.2 The trajectory starting from (.32, .34) for $r = 3.2$ is plotted here. The first 50 iterations have been discarded and the next 50 have been used for the plot. The larger dot is the equilibrium point $\mathbf{x}_3(3.2)$.

The stationary states of the branch $\mathbf{x}_3(r)$ are acceptable only for values of $r \geq 2$. It is exactly at $r = 2$ that $\mathbf{x}_2(2) = \mathbf{x}_3(2)$.

The matrix $F_\mathbf{x}(\mathbf{x}, r)$ evaluated along the branch $\mathbf{x}_3(r)$ gives

$$F_\mathbf{x}(\mathbf{x}_3, r) = \begin{pmatrix} 0 & -1 \\ r-2 & 1 \end{pmatrix}.$$

The eigenvalues of this matrix are

$$\lambda_1 = \frac{1 + \sqrt{9 - 4r}}{2}, \qquad \lambda_2 = \frac{1 - \sqrt{9 - 4r}}{2}.$$

Both eigenvalues are real, positive and smaller than 1 for $r \in (2, 2.25)$.
Hence, at $r = 2$ the attracting character of the equilibria of the second branch is passed on to the equilibria of the third branch .

The eigenvalues become complex for $r > 2.25$ and their modulus is $\sqrt{r - 2}$. They are inside the unit circle in the complex plane for $r \in (2.25, 3)$. At $r = 3$ the eigenvalues are

$$\lambda_1 = \frac{1 + \mathbf{i}\sqrt{3}}{2}, \quad \lambda_2 = \frac{1 - \mathbf{i}\sqrt{3}}{2}.$$

The eigenvalues cross the unit circle as r crosses 3 and

$$\frac{d}{dr}\sqrt{r-2} > 0 .$$

Moreover, the argument of the complex number λ_1 is

$$\theta = \mathrm{ArcTan}\sqrt{3} ,$$

which implies $\theta = \frac{\pi}{3}$ and $m = 1, n = 6$ (see Chapter 5).

Therefore, at $r = 3$ we have Hopf bifurcation. The orbits close to $(r^{-1}, 1 - 2r^{-1})$ for $r > 3$ are attracted by a periodic orbit of period 6 which is itself a function of r. The orbit is given by (see also Fig. 7.2.2) :

$$x_1(r) = \left(\frac{r-1+\sqrt{r^2-2r-3}}{2r}, \frac{r-1-\sqrt{r^2-2r-3}}{2r} \right),$$

$$x_2(r) = \left(\frac{r-1+\sqrt{r^2-2r-3}}{2r}, \frac{1}{r} \right),$$

$$x_3(r) = \left(\frac{1}{r}, \frac{r-1+\sqrt{r^2-2r-3}}{2r} \right),$$

$$x_4(r) = \left(\frac{r-1-\sqrt{r^2-2r-3}}{2r}, \frac{r-1+\sqrt{r^2-2r-3}}{2r} \right)$$

$$x_5(r) = \left(\frac{r-1-\sqrt{r^2-2r-3}}{2r}, \frac{1}{r} \right),$$

$$x_6(r) = \left(\frac{1}{r}, \frac{r-1-\sqrt{r^2-2r-3}}{2r} \right).$$

For larger values of the control parameter the system seems to be chaotic, as suggested by Fig. 7.2.3.

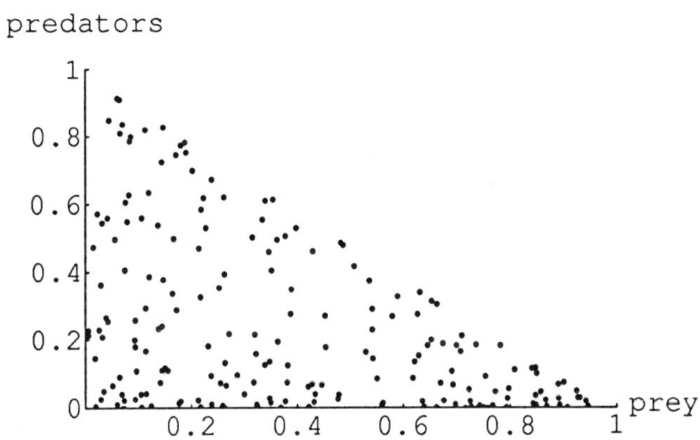

Fig. 7.2.3 The orbit starting at (.4, .2) for $r = 3.9$. 250 iterations have been discarded and the next 200 have been plotted.

Section 3. The Lorenz Model of Atmospheric Behavior

As mentioned in Chapter 1, E.N. Lorenz [Lorenz, 1963] proposed a first order system of differential equations as a crude approximation model of weather patterns. The upper atmosphere was imagined as an infinite uniform layer of depth H, with the temperature difference between the lower and the upper surface maintained at a constant value ΔT. Denoting by x, y and z quantities respectively proportional to the intensity of the convective motion between the two surfaces, to the temperature difference between the ascending and descending current, and to the deviation of the temperature fall from its linear profile , Lorenz arrived at the system

$$(7.3.1) \qquad \begin{cases} x' = -ax + ay \\ y' = -xz + rx - y \\ z' = xy - bz . \end{cases}$$

As in Example 5.4.1 we change the variable z to $u = z - a - r$. The qualitative behavior is not affected by this change and $u' = z'$. We obtain

$$(7.3.2) \qquad \begin{cases} x' = -ax + ay \\ y' = -xu - ax - y \\ u' = xy - bu - ab - br . \end{cases}$$

The one-step Euler discretization of this system with $\Delta t = h$ gives

$$(7.3.3) \qquad \begin{cases} x_{n+1} = x_n + ah(y_n - x_n) \\ y_{n+1} = y_n + h(-x_n u_n - ax_n - y_n) \\ u_{n+1} = u_n + h(x_n y_n - bu_n - b(a+r)) . \end{cases}$$

In the analysis of (7.3.3) we can use the ideas developed so far by considering it as a discrete dynamical system in $\mathbf{R}^3$. The system is governed by the function F defined by

$$F(\mathbf{x}, \mathbf{a}) = (x + ah(y - x),\ y + h(-xu - ax - y),\ u + h(xy - bu - b(a + r))).$$

We see that $\mathbf{x}_0 = (0, 0, -a - r)$ is an equilibrium point of (7.3.3). The other equilibrium points are found by setting

$$(7.3.4) \qquad x_{n+1} = x_n,\ y_{n+1} = y_n,\ u_{n+1} = u_n .$$

From the first equation we obtain $x_n = y_n$. Substituting into the second and third equation and dropping the subscripts we arrive at the system

$$(7.3.5)\qquad \begin{cases} xu = -ax - x \\ x^2 = bu + b(a + r)\,. \end{cases}$$

The first equality implies $u = -(a+1)$. Thus $x^2 = b(r-1)$ which requires $r > 1$ ($r = 1$ gives again $x = y = u + a + r = 0$) and we find two more stationary states

$$\mathbf{x}_1 = (\sqrt{b(r-1)}\,,\ \sqrt{b(r-1)},\ -a-1)\,,\ \mathbf{x}_2 = (-\sqrt{b(r-1)}\,,\ -\sqrt{b(r-1)}\,,\ -a-1)\,.$$

The derivative of F is the matrix ($\mathbf{a} = (a, h)$)

$$F_{\mathbf{x}}(\mathbf{x}, \mathbf{a}) = \begin{pmatrix} 1-ha & ha & 0 \\ -hu-ha & 1-h & -hx \\ hy & hx & 1-bh \end{pmatrix}.$$

Hence,

$$F_{\mathbf{x}}(\mathbf{x}_0, \mathbf{a}) = \begin{pmatrix} 1-ha & ha & 0 \\ hr & 1-h & 0 \\ 0 & 0 & 1-bh \end{pmatrix}$$

and we have $F_{\mathbf{x}}(\mathbf{x}_0, \mathbf{a}) = I + hG$, where

$$G = \begin{pmatrix} -a & a & 0 \\ r & -1 & 0 \\ 0 & 0 & -b \end{pmatrix}.$$

We are interested in finding the roots of $\det(\lambda I - I - hG) = 0$. Setting $\lambda - 1 = th$ we obtain $\det(thI - hG) = 0$ or, equivalently $\det(tI - G) = 0$. This gives the third degree equation in t

$$(t + b)[(t + a)(t + 1) - ar] = 0$$

whose solutions are

$$t_1 = -b,\quad t_2 = \frac{1}{2}[-(a + 1) - \sqrt{(a-1)^2 + 4ar}\,],\quad t_3 = \frac{1}{2}[-(a + 1) + \sqrt{(a-1)^2 + 4ar}\,]$$

Consequently,

$$(7.3.6)\qquad \begin{aligned} \lambda_1 &= 1 - bh \\ \lambda_2 &= 1 - \frac{h}{2}\left(a + 1 + \sqrt{(a-1)^2 + 4ar}\right) \\ \lambda_3 &= 1 - \frac{h}{2}\left(a + 1 - \sqrt{(a-1)^2 + 4ar}\right). \end{aligned}$$

Since the time step is very small , λ_1 and λ_2 can be considered positive and smaller than 1. The last eigenvalue λ_3 is also positive and it is smaller than 1 if

$$a+1 \geq \sqrt{(a-1)^2 + 4ar}\ ,$$

which is true for $r \leq 1$.

Hence, $\mathbf{x}_0$ is an attractor if $r < 1$, and it becomes a saddle point when $r > 1$. Close to $\mathbf{x}_0$ the stable and unstable manifold (when $r > 1$) are respectively two-dimensional and one-dimensional.

At $r = 1$ we obtain $t_1 = -b$, $t_2 = -(a + 1)$ and $t_3 = 0$. Thus

$$(7.3.7)\qquad \begin{aligned} \lambda_1 &= 1 - bh \\ \lambda_2 &= 1 - h(a+1) \\ \lambda_3 &= 1\,. \end{aligned}$$

Let us examine now the stability of the two additional equilibria, which arise when r crosses the value 1. The two stationary states are

$$\mathbf{x}_1 = (\alpha, \alpha, -a-1) \text{ and } \mathbf{x}_2 = (-\alpha, -\alpha, -a-1) \text{ with } \alpha = \sqrt{b(r-1)}\ .$$

A little bit of ingenuity will help us in simplifying the computation, just as we have done at $\mathbf{x}_0$. Notice that $F_{\mathbf{x}}(\mathbf{x}_1) = I + hG_{\mathbf{x}}(\mathbf{x}_1, a)$, where

$$G_{\mathbf{x}}(\mathbf{x}_1, a) = \begin{pmatrix} -a & a & 0 \\ 1 & -1 & -\alpha \\ \alpha & \alpha & -b \end{pmatrix}.$$

We are interested in finding the roots of $\det(\lambda I - I - hG_x(\mathbf{x}_1, a)) = 0$. Setting $\lambda - 1 = th$ we obtain $\det(thI - hG_x(\mathbf{x}_1, a)) = 0$, or equivalently $\det(tI - G_x(\mathbf{x}_1), a) = 0$, which gives the third degree equation

$$t^3 + t^2(a + b + 1) + t(ab + br) + 2ab(r - 1) = 0 . \tag{7.3.8}$$

As r approaches 1, the equilibrium points $\mathbf{x}_1$ and $\mathbf{x}_2$ approach $\mathbf{x}_0$ and the solutions of this cubic equation for $r = 1$ are again $t_1 = - b$, $t_2 = - (a+1)$ and $t_3 = 0$.

We now describe what is happening to the three roots as r moves past 1. Remember that the roots are continuous functions of r and that a cubic equation has always one real root, while the other two could be real or complex conjugate. To visualize the behavior of the three roots let us choose

$$a = 10 \text{ and } b = \frac{8}{3},$$

and plot the function

$$f(t, r) = t^3 + \frac{41}{3} t^2 + \frac{8}{3} t(10 + r) + \frac{160}{3} (r - 1)$$

for $r = 1$ and $r = 1.2$. From Fig. 7.3.1 we see that as r increases past 1 the root $t_2(r)$ which is the furthest from the origin, moves to the left, while $t_1(r)$ and $t_3(r)$ move towards each other. This behavior does not change as long as $a > b + 1$ and $b > 0$.

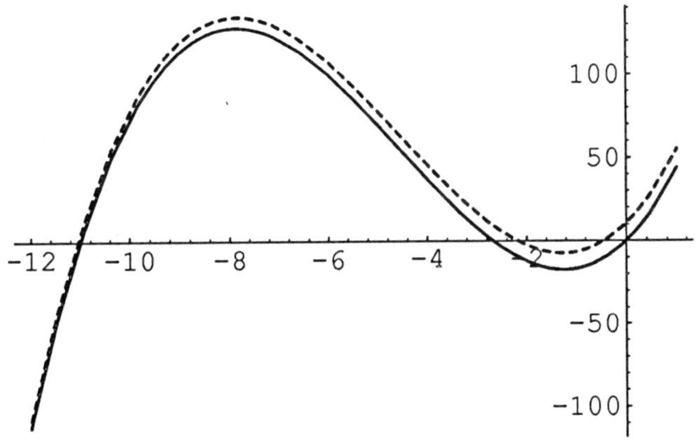

Fig. 7.3.1 The figure shows the location of the roots of the cubic equation (7.3.8) for $a = 10$, $b = 8/3$, $r = 1$, and $r = 1.2$ (dashed).

Therefore, let us imagine the three roots as points in the complex plane, with the real part plotted along the horizontal axis and the imaginary part plotted along the vertical axis. The root $t_2(r)$ is always real. The other two are real for r close to 1 and they get closer to each other as r increases.

Past the value of r for which $t_1(r)$ and $t_3(r)$ coincide, the two roots become complex conjugate with negative real part, and they start moving towards the right hand side of the imaginary axis in the complex plane.

The equilibrium point $\mathbf{x}_1$ (and $\mathbf{x}_2$) becomes unstable after the two roots cross the imaginary axis since, after that, we obtain $t_1 = c + \mathbf{i}d$, $t_3 = c - \mathbf{i}d$ with $c > 0$. Hence

$$\lambda_1 = 1 + hc + \mathbf{i}hd, \quad \lambda_2 = 1 + hc - \mathbf{i}hd$$

whose modulus is larger than 1.

At the value of r when the two roots are on the imaginary axis, our equation must have two complex conjugate solutions of the form $t = \pm\, \mathbf{i}\beta$. Substituting into (7.3.8) we get

$$\text{(7.3.9)} \qquad -\mathbf{i}\beta^3 - \beta^2(a+b+1) + \mathbf{i}\beta(ab+br) + 2ab(r-1) = 0.$$

Setting equal 0 the real and imaginary part of this expression we obtain

$$\beta^2 = ab + br \quad \text{and} \quad \beta^2 = \frac{2ab(r-1)}{a+b+1}.$$

Hence,

$$a+r = \frac{2a(r-1)}{a+b+1}$$

which gives

$$r_2 = a\frac{a+b+3}{a-b-1}.$$

Since $r > 1$, we must have $a > b+1$. Thus the stationary state $\mathbf{x}_1$ and $\mathbf{x}_2$ are unstable for $r > r_2$.

With the values $a = 10$ and $b = \frac{8}{3}$ proposed by Lorenz we obtain that the two equilibria $\mathbf{x}_1$ and $\mathbf{x}_2$ are unstable for $r > 24.74$. The analysis of the system, after selecting the slightly supercritical value $r = 28$, suggests that the dynamical process is chaotic. The point $\mathbf{x}_0$ is a saddle with two stable and one unstable direction, while the two equilibria $\mathbf{x}_1, \mathbf{x}_2$ are saddles with one stable and two unstable directions.

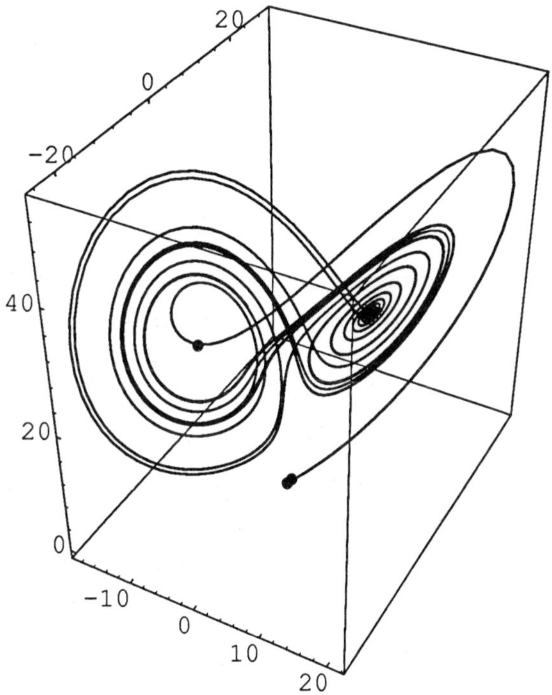

Fig. 7.3.2 A trajectory of the one-step Euler discretization of the Lorenz map for r = 28, a = 10, b = 8/3.

With $a = 10, b = 8/3$ and $r = 28$ the ellipsoid

$$10x^2 + y^2 + \frac{8}{3}(u + 19)^2 = \frac{2888}{3}$$

plays a determinant role in the dynamics.

The three equilibria are sitting on the surface of the ellipsoid. From $F = I + hG$ we derive that $G(\mathbf{x})\cdot\mathbf{x} = 0$ on the ellipsoid, while it is positive inside and negative outside. Following the ideas of Theorem 5.4.1 we know that for every real number $R > 0$ we can determine h so that every orbit starting at a point $\mathbf{x}$, $\|\mathbf{x}\| \leq R$, will remain inside the disk centered at the origin and with radius R. The orbits switch infinitely often between the region of $\mathbf{R}^3$ bounded by the ellipsoid and its complement.

In the original x, y, z variables the equilibrium point $\mathbf{x}_0$ becomes the origin and the ellipsoid takes the form

$$(7.3.10) \qquad 10x^2 + y^2 + \frac{8}{3}(z-19)^2 = \frac{2888}{3}.$$

Fig 7.3.2 shows the trajectory of an orbit starting from a point close to the origin, along the unstable manifold of the origin, and located on the surface of the ellipsoid. The time step is 0.01. The chaotic behavior of the system seems to be emerging.

Extensive numerical investigations have shown that there are orbits leaving the origin from points along the unstable manifold and arriving at one of the two other equilibria along their stable manifold. The presence of these heteroclinic orbits seems to play an important role in the chaotic behavior of the system. The conclusion suggested by this analysis is the instability and the chaotic behavior of weather patterns at least for certain values of the control parameters.

Section 4. Neural Networks

The Neural Network model presented in the first chapter is due to J.J. Hopfield [Hopfield, 1982] , and in vector notation takes the form

$$(7.4.1) \qquad \mathbf{x}_{n+1} = (I - C)\mathbf{x}_n + TF(\mathbf{x}_n) + \mathbf{x}_I \, .$$

The entry x_i of $\mathbf{x} = (x_1, ..., x_q)$ represents the potential of neuron i, i = 1, 2, ..., q. The matrix C is diagonal

$$C = \begin{pmatrix} c_1 & 0 & 0 & .... & 0 & 0 \\ 0 & c_2 & 0 & .... & 0 & 0 \\ ... & ... & ... & ... & ... & ... \\ 0 & 0 & 0 & ... & 0 & c_q \end{pmatrix}$$

and $c_i \in [0, 1]$ accounts for the leakage current of neuron i, i = 1, 2, ..., q.
The matrix $T = \{t_{ij}, q^2\}$ is the connectivity matrix, with t_{ij} denoting the action of neuron j on neuron i. This action can be excitatory ($t_{ij} > 0$), or inhibitory ($t_{ij} < 0$). The vector function F is the neuron response function $F(\mathbf{x}) = (f_1(x_1), f_2(x_2), ..., f_q(x_q))$. We can assume that all f_i's are the same and they have the following basic form

$$f(x) = \begin{cases} 0 & \text{if } x \le 0 \\ 1 & \text{if } x \ge 1 \\ x & \text{if } 0 < x < 1 \, . \end{cases}$$

The vector $\mathbf{x}_I$ represents the initial signal received by the net .

Typically, but not always, the neurons of the net are divided into three groups: **the output layer, the hidden layer and the input layer**. The neurons of the input layer are the ones which receive the signal, and the neurons of the output layer provide the answer to the problem posed by the signal. This can be a recognition problem, like identifying among a list of voices which one is represented by the signal just received. The incoming signal does not affect the hidden and the output layers directly. Therefore, if we divide the components of the vector $\mathbf{x}$ into three groups, corresponding to the three different layers, we obtain that the components of $\mathbf{x}_I$ corresponding to the first two groups are zero.

Once the signal is received, the dynamical process starts from the state in which the net is at that instant. We imagine that the signal remains active during the entire process. The initial state $\mathbf{x}_0$ of the net is usually unknown. From this state the net produces

$$\mathbf{x}_1 = (I - C)\mathbf{x}_0 + TF(\mathbf{x}_0) + \mathbf{x}_I\,,\ \mathbf{x}_2 = (I - C)\mathbf{x}_1 + TF(\mathbf{x}_1) + \mathbf{x}_I \dots .$$

The process is considered completed when the net reaches an equilibrium point and, consequently, the potential of each neuron is fixed. This equilibrium point represents the answer of the net to the problem posed by the incoming signal.

In a typical situation the net is designed to deal with finitely many inputs $\mathbf{x}_I$, $I = 1, 2, \dots, N$. Unfortunately they may be received imperfectly for several reasons, like noise, distortion, or partial occlusion. We would like that to each perfect signal corresponds the same equilibrium regardless of the initial state $\mathbf{x}_0$ of the net, and that equilibria corresponding to different perfect signals not only are different equilibria, but are also as far away as possible from each other. We would also like that a minor distortion (or occlusion) of the perfect signal generates an equilibrium point which is close enough to the one produced by the perfect signal that the net will give the same answer.

Notice that we don't need the components of the input and of the hidden layer to have reached an equilibrium state, as long as the ones of the output layer have done so, and their state provides the desired answer. This fact, however, is a mixed blessing since it makes it easier for the output layer to be the same even though the input signals are coming from different objects.

We obviously do not have time here to address all these questions. We will simply analyze some fundamental properties of the system defined by (7.4.1) on the basis of the theory developed in the previous chapters.

We first want to show that in the case when $c_i > 0$ for all $i = 1, 2, \dots, q$, every ball large enough is invariant under the dynamical system defined by (7.4.1).

In fact, each component of $F(\mathbf{x})$ is at most 1. Hence, using the norm

$$\|\mathbf{x}\|_0 = \max\{|x_i|\,;\ i = 1, 2, \dots, q\}\,,$$

we obtain $\|F(\mathbf{x})\|_0 \leq 1$. Consequently, $\|TF(\mathbf{x})\|_0 \leq \|T\|_0$ for every $\mathbf{x} \in \mathbf{R}^q$. Moreover

$$\|(I - C)\mathbf{x}\|_0 = \max\{|(1 - c_i)x_i| ; i = 1, 2, ..., q\} \leq (1 - c_0)\|\mathbf{x}\|_0 \tag{7.4.2}$$

where $1 - c_0 = \max\{1 - c_i : i = 1, 2, ..., q\}$. Thus

$$\frac{\|(I - C)\mathbf{x} + TF(\mathbf{x}) + \mathbf{x}_I\|_0}{\|\mathbf{x}\|_0} \leq 1 - c_0 + \frac{\|T\|_0}{\|\mathbf{x}\|_0} + \frac{\|\mathbf{x}_I\|_0}{\|\mathbf{x}\|_0}. \tag{7.4.3}$$

We can choose r_0 so large that $\frac{\|T\|_0}{\|\mathbf{x}\|_0} + \frac{\|\mathbf{x}_I\|_0}{\|\mathbf{x}\|_0} \leq \frac{c_0}{2}$ for every $\mathbf{x}$, $\|\mathbf{x}\|_0 \geq r_0$. Thus, under these circumstances, we have

$$\frac{\|(I - C)\mathbf{x} + TF(\mathbf{x}) + \mathbf{x}_I\|_0}{\|\mathbf{x}\|_0} \leq 1 - \frac{c_0}{2} < 1 \tag{7.4.4}$$

for all $\mathbf{x}$, $\|\mathbf{x}\|_0 \geq r_0$. Hence, (see Theorem 5.4.2) $N(r_0) \leq 1$ and the map which defines our dynamical system leaves invariant every sufficiently large ball in the $\|.\|_0$ norm.

We now show that different signals cannot produce the same equilibrium, at least in the case when we require a stationary state in all three layers. Suppose, in fact, that two signals $\mathbf{x}_I$ and $\mathbf{y}_I$ have produced two stationary states $\mathbf{x}_s$ and $\mathbf{y}_s$. From

$$\mathbf{x}_s = (I - C)\mathbf{x}_s + TF(\mathbf{x}_s) + \mathbf{x}_I, \text{ and } \quad \mathbf{y}_s = (I - C)\mathbf{y}_s + TF(\mathbf{y}_s) + \mathbf{y}_I,$$

we derive

$$C\mathbf{x}_s = TF(\mathbf{x}_s) + \mathbf{x}_I, \qquad\qquad C\mathbf{y}_s = TF(\mathbf{y}_s) + \mathbf{y}_I.$$

Therefore (we can use the Euclidean norm)

$$\|C\| \, \|\mathbf{x}_s - \mathbf{y}_s\| + \|T\| \, \|F(\mathbf{x}_s) - F(\mathbf{y}_s)\| \geq \|\mathbf{x}_I - \mathbf{y}_I\|.$$

Since $\|F(\mathbf{x}) - F(\mathbf{y})\| \leq \|\mathbf{x} - \mathbf{y}\|$ we obtain

$$\|\mathbf{x}_s - \mathbf{y}_s\| \geq \frac{\|\mathbf{x}_I - \mathbf{y}_I\|}{\|C\| + \|T\|}. \tag{7.4.5}$$

This means that different inputs cannot produce the same output, if the output is an equilibrium state involving all layers.

The existence of a unique equilibrium $\mathbf{x}_s$ corresponding to an input $\mathbf{x}_I$ regardless of the initial state $\mathbf{x}_0$ of the net, can be guaranteed when the map which defines the dynamical system is a contraction. In fact, according to Theorem 5.1.1 on contractive maps, they have a unique fixed point and every sequence of iterates converges to it. Therefore, it is important to give a sufficient condition for the map defining our dynamical system to be a contraction. This condition is $c_0 > \|T\|_0$.

We have used again the norm $\|\mathbf{x}\|_0 = \max\{|x_i|;\ i = 1, 2, ..., q\}$. In fact, letting

$$K(\mathbf{x}) = (I - C)\mathbf{x} + TF(\mathbf{x}) + \mathbf{x}_I ,$$

we have

$$K(\mathbf{x}) - K(\mathbf{y}) = (I - C)(\mathbf{x} - \mathbf{y}) + T(F(\mathbf{x}) - F(\mathbf{y})) ,$$

and we derive

$$\|K(\mathbf{x}) - K(\mathbf{y})\|_0 \leq \|I - C\|_0 \|\mathbf{x} - \mathbf{y}\|_0 + \|T\|_0 \|\mathbf{x} - \mathbf{y}\|_0 . \tag{7.4.6}$$

Since $\|I-C\|_0 \leq 1- c_0$ we see that

$$\|K(\mathbf{x}) - K(\mathbf{y})\|_0 \leq (1- c_0 + \|T\|_0)\|\mathbf{x} - \mathbf{y}\|_0 ,$$

which shows that K is a contraction whenever $c_0 > \|T\|_0$. Thus, in the presence of "large leakage and small interactions" the net, after receiving the input $\mathbf{x}_I$, will always reach the same equilibrium regardless of its initial state $\mathbf{x}_0$.

Example 7.4.1

Let $\mathbf{x} = (x, y, z)$, $c_1 = c_2 = c_3 = .05$, $\mathbf{x}_I = (0, 0, .4)$ and

$$T = \begin{pmatrix} 0 & .02 & 0 \\ -.01 & .03 & .03 \\ 0 & .01 & 0 \end{pmatrix}.$$

Then

$$\begin{cases} x_{n+1} = .95x_n + .02\, f(y_n) \\ y_{n+1} = .95y_n - .01\, f(x_n) + .03\, [f(y_n) + f(z_n)] \\ z_{n+1} = .95z_n + .01 f(y_n) + .4 . \end{cases}$$

From the initial condition $\mathbf{x}_0 = (1, -1, 1)$ the sequence of iterates , after about 300 steps, reaches the equilibrium point (0.4, 1.12, 8.2) within the prescribed degree of accuracy (5 correct decimal places). Starting from a different initial condition, namely $\mathbf{x}_0 = (0, 1, 1)$, the sequence of iterates reaches the same equilibrium point, although with fewer iterations. The result was expected since $c_0 > \|T\|_0$. The inequality implies that the map K is a contraction.

Changing the input $\mathbf{x}_I$ to (0, 0, .2), and using the same initial condition (0, 1, 1), we obtain the equilibrium point (0.4, 1.12, 4.2), which is different from the previous one. However, notice that the output component of the equilibrium point is the same, and therefore the information derived from the net would be the same. This result spells trouble if (0, 0, .4) and (0, 0, .2) represent two different objects.

Example 7.4.2

Let $c_1 = .01$, $c_2 = .02$, $c_3 = .03$, $\mathbf{x}_I = (0, 0, .2)$ and

$$T = \begin{pmatrix} .2 & .2 & 0 \\ .2 & .1 & .09 \\ 0 & -.1 & 0 \end{pmatrix}.$$

Then

$$\begin{cases} x_{n+1} &= .99x_n + .2\,[f(x_n) + f(y_n)] \\ y_{n+1} &= .98y_n + .2\,f(x_n) + .1\,f(y_n) + .09\,f(z_n) \\ z_{n+1} &= .97z_n - .1\,f(y_n) + .2\,. \end{cases}$$

The map K is not a contraction. However the system, from the initial condition (0, 1, 1), and after about 1000 iterations, reaches the equilibrium point $(40, 19.5, 3.\overline{3})$. Changing the input to (0, 0, .1), and leaving the same initial condition, gives us (40, 15, 0), a result which is different from the previous one, but with the same output.

Problems

Section 1.

7.1.1. Consider the production function

$$P(x) = bp^m \frac{x}{p^m + x^m}, \quad m > 1.$$

Show that there is a single maximum on $(0, \infty)$ and a single inflection point. Find their value and sketch the graph of the function.

7.1.2 Show that with $r = 8$, $s = 16$ and $b = 1.1 \times 10^6$, and for every $a \in (0, 1)$, the first non-zero equilibrium point x_c of the dynamical system defined by (7.1.4) is always unstable.

7.1.3 What are the values of a for which the function F of (7.1.9) has a minimum point x_m, after the largest equilibrium x_3 ? Estimate x_m numerically and check if there are any values of a such that $F(x_m, a) < x_2$. What interpretation can be given to this occurrence ?

7.1.4 Compare x_m with the maximum value of F, i.e. with $F(x_M, a)$. Is there any value of a for which $F(x_M, a) > x_m$ and $F(F(x_M, a), a) < x_2$?

Section 2.

7.2.1 A moderate harvesting activity modifies (7.2.5). Find the new equilibrium points. Is the activity in favor of the prey ?

7.2.2 What is happening to the equilibria of (7.2.5) when $c > d$?

7.2.3 Let M be a square matrix of size 2. Justify why the eigenvalues of M have negative real part if the trace of M is negative and the determinant of M is positive.

7.2.4 Let $F(x, y, r) = (rx(1 - x - y), rxy)$. Find the stationary states of the dynamical system $\mathbf{x}_{n+1} = F(\mathbf{x}_n, r)$.

7.2.5 Let $F(x, y, r) = (rx(1- x - y), rxy)$ and consider the dynamical system

$$\mathbf{x}_{n+1} = F(\mathbf{x}_n, r).$$

Verify that the equilibrium points of the branch $\mathbf{x}_s(r) = (1 - r^{-1}, 0)$ are

sinks for $r \in (1, 2)$, saddles for $r \in [2, 3)$, and repellers for $r > 3$.

7.2.6 Let F be as in the previous problem. Verify that for $r > 2.25$ the eigenvalues of the derivative of F at the equilibrium points of the branch $\mathbf{x}_3(r) = (r^{-1}, 1 - 2r^{-1})$ are complex and find their modulus.

7.2.7 Let F be as in problem (7.2.5). Verify that the derivative of the modulus of the eigenvalues of $F_{\mathbf{x}}(\mathbf{x}_3(r), r)$ at the point $r = 3$ is strictly positive.

Section 3.

7.3.1 Verify that the analysis of the stability of $\mathbf{x}_2$ brings the same conclusions as the analysis of the stability of $\mathbf{x}_1$.

7.3.2 Find the eigenvector of $F_{\mathbf{x}}(\mathbf{x}_0)$, $\mathbf{x}_0 = (0, 0, -a -r)$ corresponding to the eigenvalue larger than 1, using the parameter values $a = 10$, $b = 8/3$, and $r = 28$.

7.3.3 Verify that for the same values of the parameters, the surface of the ellipsoid (7.3.10) contains all three equilibria. Verify that the tangent plane to the ellipsoid at the origin is the xy plane.

7.3.4 Verify that the eigenvector corresponding to the eigenvalue $1 - bh$ is perpendicular to the tangent plane, while the other two eigenvectors belong to the tangent plane. Are they pair-wise orthogonal ?

7.3.5 Verify that the tangent plane to the ellipsoid (7.3.10) at $\mathbf{x}_1 = (6\sqrt{2}, 6\sqrt{2}, 27)$ has equation $30\sqrt{2}\, x + 3\sqrt{2}\, y + \frac{32}{3} z - 684 = 0$. A vector perpendicular to this plane has components $(30\sqrt{2}, 3\sqrt{2}, \frac{32}{3})$. Is this an eigenvector of the matrix $F_{\mathbf{x}}(\mathbf{x}_1)$?

Section 4.

7.4.1 Let $f(x) = \alpha x + \beta$ be the neuron response function. Show that the dynamical system (7.4.1) takes the form $\mathbf{x}_{n+1} = B\mathbf{x}_n + \mathbf{x}_J$, where B is the matrix $B = I - C + TA$, with $A = \alpha I$, and $\mathbf{x}_J$ is the vector $\mathbf{x}_J = \mathbf{x}_I + T\mathbf{b}$ with $\mathbf{b} = (\beta, \ldots, \beta)$.

7.4.2 Assume that the neuron response function is as in problem 7.4.1, but the parameters α and β are (possibly) different for each neuron, namely $f_i(x) = \alpha_i x + \beta_i$. Write the dynamical system (7.4.1) in this case.

7.4.3 When the neurons of the net are divided into the three groups :

output layer (# 1), hidden layer (# 2), and input layer (# 3),

the connectivity matrix T can be considered as composed of blocks in the following way

$$T = \begin{pmatrix} T_{11} & T_{12} & T_{13} \\ T_{21} & T_{22} & T_{23} \\ T_{31} & T_{32} & T_{33} \end{pmatrix}.$$

The T_{ij} block represent the action of the neurons of group j on the neurons of group i. The two blocks T_{13} and T_{31} are usually assumed to be zero. Can you justify this assumption? Also, what effect will have on the matrix the assumption that no neuron has any influence on itself ?

7.4.4 Assume that a neural net has 3 neurons in the output layer, 4 in the hidden layer and 2 in the input layer. Find the size of each block of T.

7.4.5 Consider a neural net with 1 neuron in the output , 2 in the hidden and 1 in the input layer. Assume that $c_i = 1$ for $i = 1, 2, 3, 4$. Also assume that the entries of the connectivity matrix T are as follows : $t_{12} = t_{13} = t_{23} = t_{34} = t_{43} = 1$, and all other entries are 0. Finally, let the neuron response function be the same for all neurons and be defined by

$$f(x) = \begin{cases} 0 & \text{if } x \leq 3/4 \\ 4x-3 & \text{if } 3/4 < x < 1 \\ 1 & \text{if } x \geq 1 . \end{cases}$$

Let

$$\mathbf{x}_I = (0, 0, 0, \tfrac{1}{2}) \text{ and let } \mathbf{x}_0 = (1, 0, 1, \tfrac{1}{2}).$$

Verify that the state of the output neuron is constant, while the states of the three remaining neurons are periodic of period 2.

References

[Banks et al., 1992], Banks J., Brooks, J. , Cairns, G., Davis, G., Stacey, P. : *On Devaney's definition of Chaos,* The American Mathematical Monthly, Vol. **99**, No.4 (1992), 332-334.

[Birkhoff, 1920], Birkhoff, G.D. : Dynamical Systems, American Mathematical Society, Colloquium Publications, Vol. 9, 1920.

[Collet-Eckmann, 1980], Collet P., Eckmann J. P. : Iterated Maps on the Interval as Dynamical Systems, Birkhäuser, Boston, 1980.

[Devaney, 1987], Devaney R. L. : An Introduction to Chaotic Dynamical Systems, Addison-Wesley, 1987.

[Fatou, 1926], Fatou P. : *Sur l'iteration des fonctionnes transcendantes entières,* Acta Math, **47** (1926), 337-370.

[Feigenbaum, 1983], Feigenbaum M. J. : *Universal behavior in nonlinear systems.* In Nonlinear Dynamics and Turbulence, Barenblatt, G.I., Iooss, G., Joseph, D., Editors, Pitman Advance Publishing Company, 1983.

[Gearhart-Martelli, 1990], Gearhart W., Martelli M. : *A blood cell population model, dynamical diseases, and chaos.* UMAP **11** 4 (1990), 311-338

[Gleick, 1987], Gleick J. : Chaos, Making a New Science. Vicking Penguin Inc., 1987.

[Grassberger-Procaccia, 1983], Grassberger P., Procaccia I. : *Measuring the strangeness of strange attractors,* Physica **9**D (1983), 189-208.

[Hénon, 1976], Hénon M. : *A two-dimensional mapping with a strange attractor,* Comm. Math. Phys. **50** (1976), 69-78.

[Hopfield, 1982], Hopfield J.J. : *Neural networks and physical systems with emergent collective computational abilities*, Proc. Natl. Acad. Sci. USA, Vol. **79** (1982), 2554-2558

[Julia, 1918], Julia G. : *Mémoirs sur l ' iteration des functions rationelles*, J. de Math. Ser. **7**, 4 (1918), 47-245.

[Lasota, 1977], Lasota A.: *Ergodic problems in Biology*, Société Mathématique de France, Astérisque, **50** (1977), 239-250.

[Lauwerier, 1986a], Lauwerier H.A. : *One-dimensional iterative maps*, In Chaos, Holden A.V. Editor, Princeton University Press, 1986, 39-58.

[Lauwerier, 1986b], Lauwerier H.A. : *Two-dimensional iterative maps*, In Chaos, Holden A.V. Editor, Princeton University Press, 1986, 58-97.

[Li-Yorke, 1975], Li T., Yorke J.A. : *Period three implies chaos*, The American Mathematical Monthly, **82** (1975), 985-992.

[Lorenz, 1963], Lorenz E.N. : *Deterministic nonperiodic flow*, J. Atmos. Sc., **20** (1963), 130-141.

[Marotto, 1979], Marotto F. R. : *Perturbation of stable and chaotic difference equations*, J. Math. An. Appl. **72** (1979), 716-729.

[May, 1976], May, R. M. : *Simple mathematical models with very complicated dynamics.* Nature, Lond. **261** (1976), 459-467.

[May, 1979], May, R. M. : *Bifurcations and dynamic complexity in ecological systems.* Ann. N.Y. Acad. Sci. **316** (1979), 517-529

[Sarkovskii, 1964], Sarkovskii, A.N. : *Coexistence of cycles of a continuous map of a line into itself*, Ukr. Mat. Z. **16** (1964), 61-71.

[Stewart, 1989], Stewart, I. : Does God play Dice ? The Mathematics of Chaos. Blackwell, 1989.

[Volterra, 1931], Volterra, V. : Leçons sur la théorie mathématique de la lutte pour la vie, Paris, 1931.

Some Computer Programs and Related Books

Models , by Lapidus M., Lascaux Graphics, Tucson, AZ. , 1-800-338-0993. Available for Macintosh and for IBM compatibles.

Chaos, Mathematics for the 21st Century, by Walh B. of Dynamic Software, P.O. Box 7534, Santa Cruz, CA 95061. Available

dstool: Computer Assisted Exploration of Dynamical Systems, by Back A., Guckenheimer J., Myers M., Wicklin F. and Worfolk P., Center for Applied Mathematics, Cornell University, Ithaca, NY 14853

Mathematica in Action, by Wagon S. , W.H. Freeman & Company, Inc. , New York, 1991.

Exploring Mathematics with Mathematica, by Grey T. and Glynn J. , Addison-Wesley Advanced Book Program, Redwood City, CA, 1991.

Programming in Mathematica, by Maeder R., Addison-Wesley Advanced Book Program, Redwood City, CA, 1990

Mathematica, by Wolfram S., Addison-Wesley Advanced Book Program, Redwood City, CA, 2nd ed. 1991.

Index

E

F

G

H

I

L

M